PENGUIN BOOKS

# HERE COMES EVERYBODY

Clay Shirky writes, teaches and consults on the social and economic effects of the internet, especially on places where our social and technological networks overlap. His goal is to describe the intersection of social tools and social life, helping people both to understand what's happening around them, and how tools could be designed that better support social activity. A professor at NYU's Interactive Telecommunications Program, he has consulted for Nokia, Procter and Gamble, News Corp., the BBC, the US Navy, and Lego. Over the years, his writings have appeared in *The New York Times*, the *Wall Street Journal*, the *Harvard Business Review*, *Wired* and *IEEE Computer*. Pivotal articles include 'Exiting Deanspace', an analysis of Howard Dean's loss of the US Democratic Nomination in 2004, and how his web campaign may actually have contributed to the loss, and 'Power Laws, Weblogs and Inequality', about the ways that the social dynamics of online communication tend to create giant imbalances of attention. A regular keynote speaker at tech conferences, he has never believed that technology is an end to itself; rather it is our use of technology that matters.

# HERE COMES EVERYBODY

## HOW CHANGE HAPPENS WHEN
## PEOPLE COME TOGETHER

## CLAY SHIRKY

PENGUIN BOOKS

*For Almaz*

PENGUIN BOOKS

Published by Penguin Group
Penguin Books Ltd, 80 Strand, London WC2R ORL, England
Penguin Group (USA) Inc., 375 Hudson Street, New York, New York 10014, USA
Penguin Group (Canada), 90 Eglinton Avenue East, Suite 700, Toronto, Ontario, Canada M4P 2Y3
(a division of Pearson Penguin Canada Inc.)
Penguin Ireland, 25 St Stephen's Green, Dublin 2, Ireland (a division of Penguin Books Ltd)
Penguin Group (Australia), 250 Camberwell Road, Camberwell, Victoria 3124, Australia
(a division of Pearson Australia Group Pty Ltd)
Penguin Books India Pvt Ltd, 11 Community Centre, Panchsheel Park,
New Delhi – 110 017, India
Penguin Group (NZ), 67 Apollo Drive, Rosedale, North Shore 0632, New Zealand
(a division of Pearson New Zealand Ltd)
Penguin Books (South Africa) (Pty) Ltd, 24 Sturdee Avenue, Rosebank,
Johannesburg 2196, South Africa

Penguin Books Ltd, Registered Offices: 80 Strand, London WC2R ORL, England

www.penguin.com

First published in the United States of America by The Penguin Press, a member of Penguin
Group (USA) Inc. 2008
First published in Great Britain by Allen Lane 2008
Published in Penguin Books 2009
006

Printed in England by Clays Ltd, St Ives plc

978-0-141-03062-3

www.greenpenguin.co.uk

**MIX**
Paper from
responsible sources
FSC   **FSC™ C018179**
www.fsc.org

Penguin Books is committed to a sustainable
future for our business, our readers and our planet.
This book is made from Forest Stewardship
Council™ certified paper.

# CONTENTS

CHAPTER 1 | It Takes a Village to Find a Phone 1

CHAPTER 2 | Sharing Anchors Community 25

CHAPTER 3 | Everyone Is a Media Outlet 55

CHAPTER 4 | Publish, Then Filter 81

CHAPTER 5 | Personal Motivation Meets Collaborative Production 109

CHAPTER 6 | Collective Action and Institutional Challenges 143

CHAPTER 7 | Faster and Faster 161

CHAPTER 8 | Solving Social Dilemmas 188

CHAPTER 9 | Fitting Our Tools to a Small World 212

CHAPTER 10 | Failure for Free  233

CHAPTER 11 | Promise, Tool, Bargain  260

EPILOGUE  293

ACKNOWLEDGMENTS  322

BIBLIOGRAPHY  325

INDEX  337

# CHAPTER 1

# IT TAKES A VILLAGE TO FIND A PHONE

On an afternoon in late May 2006 a woman named Ivanna left her phone in the backseat of a New York City cab. No surprise there; hundreds of phones a year show up in the New York Taxi and Limousine Commission's offices, and more than that are actually lost, since some unknown number are simply taken by the next passenger. That was the fate of Ivanna's phone, a fairly expensive multifunction version called a Sidekick, which came with a screen, keyboard, and built-in camera. Sadly for her, the Sidekick was the sole repository of much of the information for her upcoming wedding, from contact information for the catering company to the guest list.

When she realized what she'd done, Ivanna asked Evan Guttman, a friend who worked as a programmer in the financial industry, to offer a reward for its return, via an e-mail message that would show up on the phone. Getting no response after a couple of days, she shelled out more than $300 to buy a new one. Ivanna's phone company had stored copies of her information on its servers and transferred it to her new phone. Once the information had been transferred to her new phone, she discovered that her original one had ended up in the hands of a girl in Queens. Ivanna knew this because the girl was using it to take pictures of herself and her friends and e-mail them around; the photos taken on her old phone had also been transferred to her new one. Ivanna and Evan couldn't be sure who had taken the phone from the cab, but they knew who had it now, or rather they had her picture and her e-mail address, Sashacristal8905@aol.com (since disabled, for reasons that will become apparent).

Evan immediately e-mailed Sasha, explaining the situation and asking for the phone back. Sasha replied that she wasn't stupid enough to return it, a view punctuated with racial invective, saying that Evan's "white ass" didn't deserve it back. (She inferred Evan and Ivanna's race from pictures on the phone; Sasha is Hispanic.) The back-and-forth went on for some time. During the conversation Sasha said her brother had found it in a cab and given it to her; Evan continued to ask for it back, on the grounds that Sasha knew who its rightful owner was. Sasha finally wrote that she and her boyfriend would meet Evan, saying, in the spelling-challenged manner of casual e-mails, "i got ball this is my adress 108 20 37 av corona come n do it iam give u the sidekick so I can hit you wit it."

Evan declined to go to the listed address, both because he assumed it was fake (it was) and because of the threatened violence. Instead, he decided to take the story public. He created a simple webpage with Sasha's photos and a brief description of the events so far, with the stated rationale of delivering a lesson on "the etiquette of returning people's lost belongings," as he put it. He titled the page StolenSidekick, added it to his personal website at EvanWasHere.com, and began telling his friends about what had happened.

The original page went up on June 6, and in the first few hours it was up, Evan's friends and their friends forwarded it around the internet, attracting a growing amount of attention. Evan first updated the page later that day, noting that his friends had done some online detective work and had found a page on MySpace, the social networking website, that had photos of Sasha and a man they surmised was her boyfriend. Evan's second update provided more background on how the phone was lost and on who had it now. His third update, later that afternoon, reported that an officer from the NYPD had seen the story and had written explaining how to file a claim with the police.

That evening, two things happened. First, a man named Luis sent Evan mail, saying he was Sasha's brother and a member of the Military Police. He said that Sasha had bought the phone from a cabbie. (This story, as Evan pointed out on the webpage, directly contradicted Sasha's earlier account of her brother finding the phone.) Luis also told Evan to stop harassing Sasha, hinting violence if Evan didn't lay off. The other event that evening was that Evan's story appeared on Digg. Digg is a collaborative news website; users suggest stories, and other users rate them thumbs up or thumbs down. The Digg front page, like all newspaper front pages, is made up of stories that are both timely and important, except on Digg timeliness is measured by how recently a story was added, and importance is measured by user votes rather than by the judgment of editors. The front page of Digg gets millions of readers a day, and a lot of those readers took a look at the StolenSidekick page.

The story clearly struck a nerve. Evan was getting ten e-mails *a minute* from people asking about the phone, offering encouragement, or volunteering to help. Everyone who has ever lost something feels a diffuse sense of anger at whoever found and kept it, but this time it was personal, since Evan, and everyone reading StolenSidekick, now knew who had the phone and had seen her insulting refusals to return it. When the barrier to returning something is high, we make peace with "Finders, keepers. Losers, weepers," but when returning something becomes easier, our sympathies ebb. Finding a loose bill on the street is different from finding a wallet with ID in it, and the case of the missing Sidekick was even worse than a lost wallet. Using someone's own phone to refuse to return it to them crossed some barrier of acceptability in the eyes of many following the saga, and the taunts and threats from Sasha and her friends and family only added insult to injury.

Evan, clearly energized by the response from his growing readership, continued posting a running commentary on his webpage. He wrote forty updates in ten days, accompanied by a growing frenzy of both local and national media attention. There was a lot to update: he and the people tuning in posted more MySpace profiles of Sasha, her boyfriend Gordo, and her brother. Someone reading the StolenSidekick page figured out Sasha's full name, then her address, and drove by her house, later posting the video on the Web for all to see. Members of Luis's Military Police unit wrote to inquire about allegations that an MP was threatening a civilian and promised to look into the matter.

Evan also created a bulletin board for his readers, a place online where they could communicate with one another about the attempts to recover Ivanna's phone. Or rather, he tried to create a bulletin board, but the first such service he selected simply couldn't cope with the crush of excited users all trying to log in at the same time. Seeing this, he selected a second bulletin board service, but that too crashed under the sudden shock of demand, as did the third. (These kinds of failures, sometimes called "success crises," bring to mind Yogi Berra's famous observation about a New York restaurant: "Nobody goes there anymore. It's too crowded.") He finally found a service that could accommodate the thousands of people following the Sidekick saga, and those readers settled in, discussing every aspect of the events, from general speculation about Sasha's moral compass to a forum inviting members of the military to talk about Luis, the MP, and his involvement in the events. (As is usual with these kinds of communities, much of the conversation was off-topic; the military section of the bulletin board included a conversation about whether Luis was taking sufficient care of the uniform he was wearing in the pictures Sasha had taken.)

During this period Sasha's family and friends kept communicating with Evan about the phone, offering several inconsistent stories: her mom had bought the phone from someone, Sasha didn't have the phone, she had sold the phone, she would sell him the phone back for $100. Luis announced they were going to sue for harassment; her friends wrote in with more threatening e-mail. Evan and Ivanna filed a report with the police, who classified the phone as lost rather than stolen property, meaning they would take no action. Several people in the New York City government wrote in offering to help get the complaint amended, including a police officer who shared internal NYPD paperwork and explained how the complaint should have been handled. (Possession of this paperwork almost got Evan arrested when he later tried to get the complaint reclassified.) By this point millions of readers were watching, and dozens of mainstream news outlets had covered the story. The public airing of the NYPD's refusal to treat this case as theft generated so many public complaints that the police later reversed their stand and, after dispatching two detectives to talk with Ivanna, agreed to treat the phone as stolen rather than lost.

Then on June 15 members of the NYPD arrested Sasha, a sixteen-year-old from Corona, New York, and recovered the stolen Sidekick, which they returned to its original owner, Ivanna. As Sasha's mother memorably told a reporter the day her daughter was arrested, "I never in my life thought a phone was gonna cause me so many problems." It wasn't the phone that caused the problems, though. It was the people at the other end of the phone, people who had come together around Evan's page, who found the MySpace profiles and the family's address and helped pressure the police department, all in a busy ten days, and all of it leading to Sasha's arrest. Having achieved their stated goals of publicly calling out Sasha and retrieving the phone, Evan and Ivanna declined to press charges, and Sasha was released. Ivanna's wedding went off without a hitch, and Evan, in light of his ability to gather a crowd, began getting freelance work doing PR.

*"Give me a place to stand and a lever long enough, and I will move the world."*

The loss and return of the Sidekick is a story about many things—Evan's obsessive tendencies, Ivanna's good fortune in having him for a friend, how expensive phones have gotten—but one of the themes running through the story is the power of group action, given the right tools. Despite Evan's heroic efforts, he could not have gotten the phone returned if he had been working alone. He used his existing social network to get the word out, which in turn helped him find an enormous audience for Ivanna's plight, an audience willing to do more than just read from the sidelines. This audience gave Evan remarkable leverage in dealing with Sasha, and with the NYPD, leverage he wouldn't have had without such an engaged group following along. Indeed, the nature of that engagement puts many of the visitors to Evan's webpage in a category that Dan Gillmor, a journalist and the author of *We the Media*, calls "the former audience," those people who react to, participate in, and even alter a story as it is unfolding.

Consider the story from Sasha's point of view. She's a teenager in a media-saturated culture, she gets a very expensive, very cool phone that someone found in the back of a cab, and she decides to keep it rather than try to track down the owner. This isn't the most ethical behavior in the world, but neither is it premeditated theft, and in any case, what could go wrong? She's got her friends and family backing her up, and she surmises, correctly, that Evan isn't in any hurry to come out to Corona. Given all this, the combination of stories and threats from Sasha and her friends and family should have worked. After all, the phone was expensive, but it wasn't that expensive, and it's not like $300 would buy Evan a lot of help. If what Evan wanted was to save Ivanna the price of the phone, spending more than $300 retrieving it wouldn't make any sense.

Evan wasn't in it for the money, though. He was in it to satisfy his sense of justice. Because his commitment to the task at hand was emotional rather than financial, and because he was well-off enough, he was able to invest considerably more in the recovery effort than the phone was actually worth. His decision to present those motivations in public also helped draw people in. "This is not a religious endeavor or a moral endeavor . . . . [*sic*] this is a HUMANITY endeavor," Evan wrote at one point. The story of righting a wrong is a powerful one and helped him generate the involvement of others that finally led to the recovery of the phone.

Sasha and her friends didn't just want Evan to fail—they assumed that he would fail. The threats from Luis and Gordo had a kind of "You and what army?" quality about them, because they were certain that the police weren't going to get involved. (Luis made this very point in his first message to Evan: "dont give me that bullshit about you going to the cops over a lost phone the nypd has better things to do then to worry about your friend losing her phone." [*sic*]) The turning point in Evan's quest was the moment when the police agreed to amend the complaint from "lost property" (about which they would have done nothing) to "stolen property" (which led to Sasha's arrest). The NYPD is not an easy organization to browbeat, yet days after they'd tried to close the case, there they were, sending two detectives to spend half an hour with Ivanna on the matter, then sending more officers out to Corona to collar Sasha and retrieve the Sidekick. Imagine how disorienting it must have been for Sasha to learn that the owner of the phone actually did have an army of sorts, including lawyers and cops, along with an international audience of millions.

Thanks to the Web, the cost of publishing globally has collapsed. That raw publishing capability, Evan's existing social contacts, the unusual nature of his story, and the fact that the audience could find Sasha's MySpace page all combined to create a kind of positive reinforcement of attention. People became interested in the story, and they forwarded it to friends and colleagues, who became interested in turn and forwarded it still further. This pattern of growth was both cause and effect for mainstream media getting involved—it's unlikely that *The New York Times* or CNN would have covered the story of a lost phone, but when it was wrapped in the larger story of national and even global attention, they picked it up, which led to still more visitors to Evan's site and still more media outlets tuning in. The story ended up in more than sixty newspapers and radio and TV stations and more than two hundred weblogs. From the humble beginnings of Ivanna's plight and a handful of snapshots of Sasha and her friends, the StolenSidekick page went on to get over a million viewers.

Having the attention of this audience changed the conditions for Evan's relations with the police, and he knew it. He even said in one of his updates that the function of the StolenSidekick page was to put pressure on the NYPD. It also emboldened him. When he went down to the Ninth Precinct to get the complaint upgraded from lost to stolen property, Evan was stymied by the desk officer, who told him in no uncertain terms that it was up to the NYPD to determine what was a crime and what wasn't. Evan's update later that day read, in part, "All I want to do is report a crime. This is ridiculous. Have no fear though. I have many surprises for the NYPD tomorrow. They WILL listen to me and the thousands of you who have written me and the millions of you who are reading this page." The surprise that he knew was coming was the appearance of the story in *The New York Times* the following morning. Later, when the police indicated a willingness to pursue the case, Evan posted an explicit request to the site: "I ask that EVERYONE come back to visit this page for updates to make sure that the NYPD stay true to what they said." Faced with the opacity of the NYPD bureaucracy, Evan had the information-age equivalent of being able to see through walls: he got insider advice, and he was able to walk into a confrontation with a New York City cop knowing that the story would be front-page news the following morning.

You can see Evan coming to accept his part of the bargain with his users—they would provide the attention that kept him going and made the story attractive to traditional media, and he would channel that attention, reporting on his every move. Many of the viewers of the StolenSidekick page were not just readers but operated as one-person media outlets, members of the former audience, and they discussed the situation on weblogs, on mailing lists, and on various electronic discussion groups Evan set up. He had lawyers, policemen, online detectives, journalists, and even his own ad hoc pressure group working on his behalf, without belonging to any organization responsible for providing those functions.

Evan's updates included mention of constant encouragement and offers of help from more people in the city government who thought he was getting a raw deal from the NYPD. Hours after he posted the first version of the page, an NYPD officer contacted him to explain how to file a complaint. Four days later another officer from the NYPD wrote Evan wanting to meet; when they did, the officer gave Evan copies of internal NYPD paperwork to show him the kind of form he needed to file to get it treated as a theft. Finally, when Sasha's family began threatening legal action, someone from Legalmatch .org, a legal advice site, offered to help Evan get free advice.

Obviously, much about this story is unrepeatable. It isn't a worldwide media event every time someone loses a phone. The unusualness of the story, though, throws into high relief the difference between past and present. It's unlikely that Evan could have achieved what he did even five years ago, and inconceivable that he could have achieved it ten years ago, because neither the tools he used nor the social structures he relied on were in place ten years ago. Equally obviously, much about this story depends on the angle you are viewing it from. For Ivanna, the story is mostly good. She benefited from Evan's obsessive behavior and the way it was fed by the attention he received, and she had to expend little effort to get her phone back. For Evan himself, the exhilaration of fighting for what he thought was right was balanced against the investment of time and expense. And for Sasha, of course, the story was mostly bad. Of all the telephones in all the towns in all the world, the one she got had a million people at the other end of the line.

And what about us? What about the society in which this tug-of-war was happening? For us the picture isn't so clear. The whole episode demonstrates how dramatically connected we've become to one another. It demonstrates the ways in which the information we give off about our selves, in photos and e-mails and MySpace pages and all the rest of it, has dramatically increased our social visibility and made it easier for us to find each other but also to be scrutinized in public. It demonstrates that the old limitations of media have been radically reduced, with much of the power accruing to the former audience. It demonstrates how a story can go from local to global in a heartbeat. And it demonstrates the ease and speed with which a group can be mobilized for the right kind of cause.

But who defines what kind of cause is right? Evan's ability to get help can be ascribed either to a strong sense of injustice or to a petty unwillingness to lose a fight, no matter how trivial and no matter the cost to his opponent. And for all the offensiveness of Sasha's taunting, race and class do matter. Evan is a grown-up doing work that lets him take countless hours off to work on the retrieval of a phone. Sasha is an unwed teenage mother. The recovery of the phone wasn't the only loss she suffered—Evan's bulletin board quickly became host to public messages disparaging Sasha, her boyfriend and friends, single mothers, and Puerto Ricans as a group. One conversation, headed with the subject line "[D]o something already!," noted that other people following the story had already uncovered her address, and advocated physical confrontation (though the author didn't offer to participate). Another thread, with the charming title "[W]ould you tap that?," involved discussion by the male participants as to whether Sasha was attractive enough to sleep with.

One could blame Evan for letting these kinds of racist and sexist conversations take place, but the number of people interested in talking about the stolen phone (as evidenced by the inadequacy of most software to handle the volume of users), and the standard anonymity of internet users, made the conversations effectively impossible to police. Furthermore, though Evan was clearly benefiting from having generated the attention, he was not entirely in control of it—the bargain he had crafted with his users had him performing the story they wanted to see. Had he shut down the bulletin boards or even edited the conversations, he would have been violating his half of what had quickly become a mutual expectation. (Whether he *should* have taken this step is a judgment call; the point is that once a group has come together, those kinds of issues of community control aren't simple. Any action Evan took, either letting the conversation go or stifling it, would have created complicated side effects.)

A larger question transcends the individual events. Do we want a world in which a well-off grown-up can use this kind of leverage to get a teenager arrested, as well as named and shamed on a global platform, for what was a fairly trivial infraction? The answer is yes and no. Millions of people obviously wanted to follow the story, in part because of its mix of moral and visceral struggle. Furthermore, what Sasha did was wrong, and we want misdeeds to be punished. At the same time, though, we want the punishment to fit the crime. It's easy enough to say that Sasha shouldn't have gotten off just because other people take lost property without returning it, but that logic starts to look different if we imagine that the roles were reversed. Poor people lose phones too, and the loss hits them far harder; why should Evan have been able to browbeat the NYPD into paying attention to this of all lost property?

A few years ago Evan wouldn't have been able to get the story heard either. Before the Web became ubiquitous, he wouldn't have been able to attract an audience, much less one

_OMG!_

_-That Was A Crazy Story!_

_-1 Wonder if it Really Happened?_

_- 1 Think She Could Have Save All The Harrasment if She Just Had Returned The Phone._

would not have ...int. Given how ...onsive bureau- ...g success, but ...willingness of ...want a world ...erage gets riled ...f the local po-

...ince that's the ...What happens next? The story of the lost Sidekick is an illustration of the kinds of changes—some good, some bad, most too complex to label— that are affecting the ways groups assemble and cooperate. These changes are profound because they are amplifying or extending our essential social skills, and our characteristic social failings as well.

## New Leverage for Old Behaviors

Human beings are social creatures—not occasionally or by accident but always. Sociability is one of our core capabilities, and it shows up in almost every aspect of our lives as both cause and effect. Society is not just the product of its individual members; it is also the product of its constituent groups. The aggregate relations among individuals and groups, among individuals within groups, and among groups forms a network of astonishing complexity. We have always relied on group effort for survival; even before the invention of agriculture, hunting and gathering required coordinated work and division of labor. You can see an echo of our talent for sociability in the language we have for groups; like a real-world version of the mythical seventeen Eskimo words for snow, we use incredibly rich language in describing human association. We can make refined distinctions between a corporation and a congregation, a clique and a club, a crowd and a cabal. We readily understand the difference between transitive labels like "my wife's friend's son" and "my son's friend's wife," and this relational subtlety permeates our lives. Our social nature even shows up in negation. One of the most severe punishments that can be meted out to a prisoner is solitary confinement; even in a social environment as harsh and attenuated as prison, complete removal from human contact is harsher still.

Our social life is literally primal, in the sense that chimpanzees and gorillas, our closest relatives among the primates, are also social. (Indeed, among people who design software for group use, human social instincts are sometimes jokingly referred to as the monkey mind.) But humans go further than any of our primate cousins: our groups are larger, more complex, more ordered, and longer lived, and critically, they extend beyond family ties to include categories like friends, neighbors, colleagues, and sometimes even strangers. Our social abilities are also accompanied by high individual intelligence. Even cults, the high-water mark of surrender of individuality to a group, can't hold a candle to a beehive in terms of absolute social integration; this makes us different from creatures whose sociability is more enveloping than ours.

This combination of personal smarts and social intuition makes us the undisputed champions of the animal kingdom in flexibility of collective membership. We act in concert everywhere, from tasks like organizing a birthday party (itself a surprisingly complicated exercise) to running an organization with thousands or even millions of members. This skill allows groups to tackle tasks that are bigger, more complex, more dispersed, and of longer duration than any person could tackle alone. Building an airplane or a cathedral, performing a symphony or heart surgery, raising a barn or razing a fortress, all require the distribution, specialization, and coordination of many tasks among many individuals, sometimes unfolding over years or decades and sometimes spanning continents.

We are so natively good at group effort that we often factor groups out of our thinking about the world. Many jobs that we regard as the province of a single mind actually require a crowd. Michelangelo had assistants paint part of the Sistine Chapel ceiling. Thomas Edison, who had over a thousand patents in his name, managed a staff of two dozen. Even writing a book, a famously solitary pursuit, involves the work of editors, publishers, and designers; getting this particular book into your hands involved additional coordination among printers, warehouse managers, truck drivers, and a host of others in the network between me and you. Even if we exclude groups that are just labels for shared characteristics (tall people, redheads), almost everyone belongs to multiple groups based on family, friends, work, religious affiliation, on and on. The centrality of group effort to human life means that anything that changes the way groups function will have profound ramifications for everything from commerce and government to media and religion.

One obvious lesson is that new technology enables new kinds of group-forming. The tools Evan Guttman availed himself of were quite simple—the phone itself, e-mail, a webpage, a discussion forum—but without them the phone would have stayed lost. Every step of the way he was able to escape the usual limitations of private life and to avail himself of capabilities previously reserved for professionals: he used his site to tell the story without being a journalist, he found Sasha's information without being a detective, and so on. The transfer of these capabilities from various professional classes to the general public is epochal, built on what the publisher Tim O'Reilly calls "an architecture of participation."

When we change the way we communicate, we change society. The tools that a society uses to create and maintain itself are as central to human life as a hive is to bee life. Though the hive is not part of any individual bee, it is part of the colony, both shaped by and shaping the lives of its inhabitants. The hive is a social device, a piece of bee information technology that provides a platform, literally, for the communication and coordination that keeps the colony viable. Individual bees can't be understood separately from the colony or from their shared, co-created environment. So it is with human networks; bees make hives, we make mobile phones.

But mere tools aren't enough. The tools are simply a way of channeling existing motivation. Evan was driven, resourceful, and unfortunately for Sasha, very angry. Had he presented his mission in completely self-interested terms ("Help my friend save $300!") or in unattainably general ones ("Let's fight theft everywhere!"), the tools he chose wouldn't have mattered. What he did was to work out a message framed in big enough terms to inspire interest, yet achievable enough to inspire confidence. (This sweet spot is what Eric Raymond, the theorist of open source software, calls "a plausible promise.") Without a plausible promise, all the technology in the world would be nothing more than all the technology in the world.

As we saw in the saga of the lost Sidekick, getting the free and ready participation of a large, distributed group with a variety of skills—detective work, legal advice, insider information from the police to the army—has gone from impossible to simple. There are many small reasons for this, both technological and social, but they all add up to one big change: forming groups has gotten a lot easier. To put it in economic terms, the costs incurred by creating a new group or joining an existing one have fallen in recent years, and not just by a little bit. They have collapsed. ("Cost" here is used in the economist's sense of anything expended—money, but also time, effort, or attention.) One of the few uncontentious tenets of economics is that people respond to incentives. If you give them more of a reason to do something, they will do more of it, and if you make it easier to do more of something they are already inclined to do, they will also do more of it.

Why do the economics matter, though? In theory, since humans have a gift for mutually beneficial cooperation, we should be able to assemble as needed to take on tasks too big for one person. If this were true, anything that required shared effort—whether policing, road construction, or garbage collection—would simply arise out of the motivations of the individual members. In practice, the difficulties of coordination prevent that from happening. (Why this is so is the subject of the next chapter.)

But there are large groups. Microsoft, the U.S. Army, and the Catholic Church are all huge, functioning institutions. The difference between an ad hoc group and a company like Microsoft is management. Rather than waiting for a group to self-assemble to create software, Microsoft manages the labor of its employees. The employees trade freedom for a paycheck, and Microsoft takes on the costs of directing and monitoring their output. In addition to the payroll, it pays for everything from communicating between senior management and the workers (one of the raisons d'être for middle management) to staffing the human resources department to buying desks and chairs. Why does Microsoft, or indeed any institution, tolerate these costs?

They tolerate them because they have to; the alternative is institutional collapse. If you want to organize the work of even dozens of individuals, you have to manage them. As organizations grow into the hundreds or thousands, you also have to manage the managers, and eventually to manage the managers' managers. Simply to exist at that size, an organization has to take on the costs of all that management. Organizations have many ways to offset those costs—Microsoft uses revenues, the army uses taxes, the church uses donations—but they cannot avoid them. In a way, every institution lives in a kind of contradiction: it exists to take advantage of group effort, but some of its resources are drained away by directing that effort. Call this the institutional dilemma—because an institution expends resources to manage resources, there is a gap between what those institutions are capable of in theory and in practice, and the larger the institution, the greater those costs.

Here's where our native talent for group action meets our new tools. Tools that provide simple ways of creating groups lead to new groups, lots of new groups, and not just more groups but more kinds of groups. We've already seen this effect in the tools that Evan used—a webpage for communicating with the world, instant messages and e-mails by the thousands among his readers, and the phone itself, increasingly capable of sending messages and pictures to groups of people, not just to a single recipient (the historical pattern of phone use).

If we're so good at social life and shared effort, what advantages are these tools creating? A revolution in human affairs is a pretty grandiose thing to attribute to a ragtag bunch of tools like e-mail and mobile phones. E-mail is nice, but how big a deal can it be in the grand scheme of things? The answer is, "Not such a big deal, considered by itself." The trick is not to consider it by itself. All the technologies we see in the story of Ivanna's phone, the phones and computers, the e-mail and instant messages, and the webpages, are manifestations of a more fundamental shift. We now have communications tools that are flexible enough to match our social capabilities, and we are witnessing the rise of new ways of coordinating action that take advantage of that change. These communications tools have been given many names, all variations on a theme: "social software," "social media," "social computing," and so on. Though there are some distinctions between these labels, the core idea is the same: we are living in the middle of a remarkable increase in our ability to share, to cooperate with one another, and to take collective action, all outside the framework of traditional institutions and organizations. Though many of these social tools were first adopted by computer scientists and workers in high-tech industries, they have spread beyond academic and corporate settings. The effects are going to be far more widespread and momentous than just recovering lost phones.

By making it easier for groups to self-assemble and for individuals to contribute to group effort without requiring formal management (and its attendant overhead), these tools have radically altered the old limits on the size, sophistication, and scope of unsupervised effort (the limits that created the institutional dilemma in the first place). They haven't removed them entirely—issues of complexity still loom large, as we will see—but the new tools enable alternate strategies for keeping that complexity under control. And as we would expect, when desire is high and costs have collapsed, the number of such groups is skyrocketing, and the kinds of effects they are having on the world are spreading.

## The Tectonic Shift

For most of modern life, our strong talents and desires for group effort have been filtered through relatively rigid institutional structures because of the complexity of managing groups. We haven't had all the groups we've wanted, we've simply had all the groups we could afford. The old limits of what unmanaged and unpaid groups can do are no longer in operation; the difficulties that kept self-assembled groups from working together are shrinking, meaning that the number and kinds of things groups can get done without financial motivation or managerial oversight are growing. The current change, in one sentence, is this: most of the barriers to group action have collapsed, and without those barriers, we are free to explore new ways of gathering together and getting things done.

George W.S. Trow, writing about the social effects of television in *Within the Context of No Context*, described a world of simultaneous continuity and discontinuity:

> Everyone knows, or ought to know, that there has
> happened under us a Tectonic Plate Shift [. . .] the
> political parties still have the same names; we still
> have a CBS, an NBC, and a *New York Times;* but we
> are not the same nation that had those things
> before.

Something similar is happening today. Most of the institutions we had last year we will have next year. In the past the hold of those institutions on public life was irreplaceable, in part because there was no alternative to managing large-scale effort. Now that there is competition to traditional institutional forms for getting things done, those institutions will continue to exist, but their purchase on modern life will weaken as novel alternatives for group action arise.

This is not to say that corporations and governments are

going to wither away. Though some of the early utopianism around new communications tools suggested that we were heading into some sort of posthierarchical paradise, that's not what's happening now, and it's not what's going to happen. None of the absolute advantages of institutions like businesses or schools or governments have disappeared. Instead, what has happened is that most of the *relative* advantages of those institutions have disappeared—relative, that is, to the direct effort of the people they represent. We can see signs of this in many places: the music industry, for one, is still reeling from the discovery that the reproduction and distribution of music, previously a valuable service, is now something their customers can do for themselves. The Belarusian government is trying to figure out how to keep its young people from generating spontaneous political protests. The Catholic Church is facing its first prolonged challenge from self-organized lay groups in its history. But these stories and countless others aren't just about something happening to particular businesses or governments or religions. They are about something happening in the world.

Group action gives human society its particular character, and anything that changes the way groups get things done will affect society as a whole. This change will not be limited to any particular set of institutions or functions. For any given organization, the important questions are "When will the change happen?" and "What will change?" The only two answers we can rule out are never, and nothing. The ways in which any given institution will find its situation transformed will vary, but the various local changes are manifestations of a single deep source: newly capable groups are assembling, and they are working without the managerial imperative and outside the previous strictures that bounded their effectiveness. These changes will transform the world everywhere groups of people come together to accomplish something, which is to say everywhere.

# CHAPTER 2

# SHARING ANCHORS COMMUNITY

*Groups of people are complex, in ways that make those groups hard to form and hard to sustain; much of the shape of traditional institutions is a response to those difficulties. New social tools relieve some of those burdens, allowing for new kinds of group-forming, like using simple sharing to anchor the creation of new groups.*

Imagine you are standing in line with thirty-five other people, and to pass the time, the guy in front of you proposes a wager. He's willing to bet fifty dollars that no two people in line share a birthday. Would you take that bet?

If you're like most people, you wouldn't. With thirty-six people and 365 possible birthdays, it seems like there would only be about a one-in-ten chance of a match, leaving you a 90 percent chance of losing fifty dollars. In fact, you should take the bet, since you would have better than an 80 percent chance of *winning* fifty dollars. This is called the Birthday Paradox (though it's not really a paradox, just a surprise), and it illustrates some of the complexities involved in groups.

Most people get the odds of a birthday match wrong for two reasons. First, in situations involving many people, they think about themselves rather than the group. If the guy in line had asked, "What are the odds that someone in this line shares *your* birthday?" that would indeed have been about a one in ten chance, a distinctly bad bet. But in a group, other people's relationship to you isn't all that matters; instead of counting people, you need to count links between people. If you're comparing your birthday with one other person's, then there's only one comparison, which is to say only one chance in 365 of a match. If you're comparing birthdays in a group with two other people—you, Alice, and Bob, say—you might think you'd have two chances in 365, but you'd be wrong. There are three comparisons: your birthday with Alice's, yours with Bob's, and Alice's with Bob's. With four people, there are six such comparisons, half of which don't involve you at all; with five, there are ten, and so on. By the time you are at thirty-six people, there are more than six hundred pairs of birthdays. Everyone understands that the chance of any two people in a group sharing a birthday is low; what they miss is that a count of "any two people" rises much faster than the number of people themselves. This is the engine of the Birthday Paradox.

This rapidly rising number of pairs is true of any collection of things: if you have a bunch of marbles, the number of possible pairs will be set by the same math. The growing complexity gets much more wretched in social settings, however; marbles don't have opinions, but people do. As a group grows to even modest size, getting universal agreement becomes first difficult, then impossible. This quandary can be illustrated with a simple scenario. You and a friend want to go out to a movie. Before you buy the tickets, you'll have to factor in your various preferences: comedy or romance, early show or late, near work or near home. All of these will have some effect on your mutual decision, but with just two of you, getting to some acceptable outcome is fairly easy.

Now imagine that you and three friends decide to go out to a movie. This is harder, because the group's preferences are less likely to overlap neatly. Two of you love action films, two hate them; one wants the early show, three the late one, and so on. With two people, you have only one agreement to make. With four, as Birthday Paradox math tells us, you need six such agreements. Other things being equal, coordinating anything with a group of four is six times as hard as with two people,

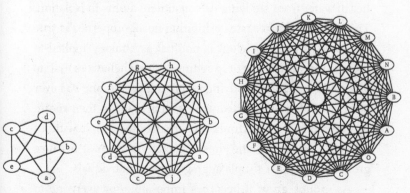

**Figure 2-1:** Three clusters, with all connections drawn. The small cluster has 5 members and 10 connections; the middle one has 10 members and 45 connections; and the large one has 15 and 105. A group's complexity grows faster than its size.

and the effect gets considerably worse as the group grows even moderately large. By the time you want to go to a movie in a group of ten, waiting for forty-five separate agreements is pretty much a lost cause. You could sit around discussing the possible choices all day, with no guarantee you'll get to an agreement at all, much less in time for the movie. Instead you'll vote or draw straws, or someone will just decide to go to a particular movie and invite everyone else along, without trying to take all possible preferences into account. These difficulties have nothing to do with friendship or movie-going specifically; they are responses to the grim logic of group complexity.

This complexity means, in the words of the physicist Philip Anderson, that "more is different." Writing in *Science* magazine in 1972, Anderson noted that aggregations of anything from atoms to people exhibit complex behavior that cannot be predicted by observing the component parts. Chemistry isn't just applied physics—you cannot understand all the properties of water from studying its constituent atoms in isolation. This pattern of aggregates exhibiting novel properties is true of people as well. Sociology is not just psychology applied to groups; individuals in group settings exhibit behaviors that no one could predict by studying single minds. No one has ever been bashful or extroverted while sitting alone in their room, no one can be a social climber or a man of the people without reference to society, and these characteristics exist because groups are not just simple aggregations of individuals.

As groups grow, it becomes impossible for everyone to interact directly with everyone else. If maintaining a connection between two people takes any effort at all, at some size that effort becomes unsustainable. You can see this phenomenon even in simple situations, such as when people clink glasses during a toast. In a small group, everyone can clink with everyone else; in a larger group, people touch glasses only with those near them. Similarly, as Fred Brooks noted in his book *The Mythical Man-Month*, adding more employees to a late project tends to make it later, because the new workers increase the costs of coordinating the group. Because this constraint is so basic, and because the problem can never be solved, only managed, every large group has to grapple with it somehow. For all of modern life, the basic solution has been to gather people together into organizations.

We use the word "organization" to mean both the state of being organized and the groups that do the organizing—"Our organization organizes the annual conference." We use one word for both because, at a certain scale, we haven't been able to get organization without organizations; the former seems to imply the latter. The typical organization is hierarchical, with workers answering to a manager, and that manager answering to a still-higher manager, and so on. The value of such hierarchies is obvious—it vastly simplifies communication among the employees. New employees need only one connection, to their boss, to get started. That's much simpler than trying to have everyone talk to everyone.

Running an organization is difficult in and of itself, no matter what its goals. Every transaction it undertakes—every contract, every agreement, every meeting—requires it to expend some limited resource: time, attention, or money. Because of these transaction costs, some sources of value are too costly to take advantage of. As a result, no institution can put all its energies into pursuing its mission; it must expend considerable effort on maintaining discipline and structure, simply to keep itself viable. Self-preservation of the institution becomes job number one, while its stated goal is relegated to number two or lower, no matter what the mission statement says. The problems inherent in managing these transaction costs are one of the basic constraints shaping institutions of all kinds.

This ability of the traditional management structure to simplify coordination helps answer one of the most famous questions in all of economics: If markets are such a good idea, why do we have organizations at all? Why can't all exchanges of value happen in the market? This question originally was posed by Ronald Coase in 1937 in his famous paper "The Nature of the Firm," wherein he also offered the first coherent explanation of the value of hierarchical organization. Coase realized that workers could simply contract with one another, selling their labor, and buying the labor of others in turn, in a market, without needing any managerial oversight. However, a completely open market for labor, reasoned Coase, would underperform labor in firms because of the transaction costs, and in particular the costs of discovering the options and making and enforcing agreements among the participating parties. The more people are involved in a given task, the more potential agreements need to be negotiated to do anything, and the greater the transaction costs, as in the movie example above.

A firm is successful when the costs of directing employee effort are lower than the potential gain from directing. It's tempting to assume that central control is better than markets for arranging all sorts of group effort. (Indeed, during the twentieth century much of the world lived under governments that made that assumption.) But there is a strong limiting factor to this directed management, and that is the cost of management itself. Richard Hackman, a Harvard professor of psychology, has studied the size and effectiveness of work groups in *Leading Teams*. Hackman tells a story about a man who ran a nonprofit whose board of directors numbered forty. When asked what he thought such a large board could accomplish, he replied, "Nothing," in a way that implied he liked it that way. Because of managerial overhead, large groups can get bogged down, and whenever transaction costs become too expensive to manage within a single organization, markets outperform firms (and central management generally).

Activities whose costs are higher than the potential value for both firms and markets simply don't happen. Here is the institutional dilemma again: because the minimum costs of being an organization in the first place are relatively high, certain activities may have some value but not enough to make them worth pursuing in any organized way. New social tools are altering this equation by lowering the costs of coordinating group action. The easiest place to see this change is in activities that are too difficult to be pursued with traditional management but that have become possible with new forms of coordination.

## How Did All Those Pictures Get There?

On the last Saturday in June, Coney Island kicks off the summer with the Mermaid Parade, a sort of hometown procession for New York City hipsters. Hundreds of people show up to march around Brooklyn's famously run-down amusement park in costumes that are equal parts extravagant and weird— a giant red octopus puppet, a flotilla of hula-hooping mermaids, a marcher sporting a bikini top made of two skulls. Thousands turn out to watch and photograph the festivities, taking pictures ranging from a couple of snapshots to dozens of high-quality photos.

A handful of these pictures end up in local newspapers, but for most of the history of the Mermaid Parade, most pictures were seen only by the people who took them and a few of their friends. The sponsor of the parade didn't provide any way for the photographers to aggregate or share their photos, and the photographers themselves didn't spontaneously organize to do so. That is the normal state of affairs. Given the complexities of group effort, hundreds of people don't spontaneously do much of any consequence, and it wouldn't have made much sense for anyone to expend the effort to identify and coordinate the photographers from the outside. A couple of years ago, however, the normal state of affairs stopped operating.

In 2005, for the first time, a hundred or so of the attendees pooled thousands of their Mermaid Parade photos and made them publicly available online. The photos came from all sorts of photographers, from amateurs with camera-phones to pros with telephoto lenses. The group was mainly populated by casual contributors—most people uploaded fewer than a dozen photos—but a handful of dedicated contributors shared more than a hundred pictures each, and one user, going by the online name czarina, shared more than two hundred photos on her own. The group pooled these photos by uploading them to a service called Flickr, giving each of the photos a free-form label called a tag. As a result, anyone can go to Flickr today, search for the tag "mermaidparade," and see the photos. This is a simple chain of events: people take pictures, people share pictures, you see pictures. It's so simple, in fact, that it's easy to overlook the substantial effort involved behind the scenes.

Flickr is the source of the sharing, but here's what Flickr did *not* do to get the sharing to happen: it didn't identify the Mermaid Parade as an interesting event, nor did it coordinate parade photographers or identify parade photographs. What it did instead was to let the users label (or tag) their photos as a way of arranging them. When two or more users adopted the same tag, those photos were automatically linked. The users were linked as well; the shared tag became a potential stepping-stone from one user to another, adding a social dimension to the simple act of viewing. The distinction between Flickr coordinating users versus helping them coordinate themselves seems minor, but it is in fact vital, as it is the only way Flickr can bear the costs involved. Consider what it would have taken for Flickr to organize hundreds of amateur mermaid photographers. Someone at Flickr HQ would have had to know about an obscure parade on the other side of the country. (Flickr is based in California.) They would have had to propose a tag for the group to use in order to assemble the uploaded photos. Finally, they would have had to communicate the chosen tag to everyone going to the parade.

This last step is especially hard. When you are trying to address a diffuse group, you are locked into the dilemma that all advertisers face: how do you reach the people you want, without having to broadcast your message to everybody?

People in the category "Potential photographer of the Mermaid Parade" aren't easy to find. Flickr couldn't have known in advance who would go to the parade. Instead, they would have to send messages out to many more people than would actually attend, in hopes of reaching the right audience, advertising to photographers, hipsters, New Yorkers, and so on, in hope of getting the tiny fraction of those groups who would actually go. Most such ads would be seen by people who weren't going to the parade, while most of the people who were going wouldn't see (or pay attention to) the ads. Given those obstacles, no business in the world would take on the job. The profit motive is little help; no one could sell enough pictures, even the skull-bikini ones, to be able to pay the photographers, much less leave any profit afterward. Likewise, no nonprofit or government agency would touch the problem; even the porkiest of pork-barrel projects isn't going to cover publicity for hula-hooping mermaids. The gap between effort and payoff is too large for any institution to span.

Yet there the photos are. Without spending any serious effort on any individual set of photos, and without doing anything to coordinate or even identify groups of photographers, Flickr has provided a platform for the users to aggregate the photos themselves.

The difference between the value of the photos and the cost of aggregation is a general one. Flickr isn't just for photos of dancing mermaids, family reunions, and the effects of that third margarita; it also hosts photos of broad public interest. Flickr provided some of the first photos of the London Transport bombings in 2005, including some taken with camera-phones by evacuees in the Underground's tunnels.

Flickr beat many traditional news outlets by providing these photos, because there were few photojournalists in the affected parts of the transport network (three separate trains on the Underground, and a bus), but many people near those parts of the transport system had camera-phones that could e-mail the pictures in. Having cameras in the hands of amateurs on the scene was better than having cameras in the hands of professionals who had to travel.

The photos that showed up after the bombings weren't just amateur replacements for traditional photojournalism; people did more than just provide evidence of the destruction and its aftermath. They photographed official notices ("All Underground services are suspended"), notes posted in schools ("Please do not inform children of the explosions"), messages of support from the rest of the world ("We love you London"), and within a day of the bombings, expressions of defiance addressed to the terrorists ("We are not afraid" and "You will fail"). Not only did Flickr host all of these images, they made them available for reuse, and bloggers writing about the bombings were able to use the Flickr photos almost immediately, creating a kind of symbiotic relationship among various social tools. The images also garnered comments on the Flickr site. A user going by Happy Dave posted an image reading "I'm OK," meant to alert his friends who had subscribed to his images on Flickr; he received dozens of comments from well-wishers in the comments. The "Do not inform the children" image generated a conversation about how to talk to kids about terrorism. The basic capabilities of tools like Flickr reverse the old order of group activity, transforming "gather, then share" into "share, then gather." People were able to connect after discovering one another through their photos.

A similar change in the broadcasting of evidence happened after the awful destruction caused by the Indian Ocean tsunami at the end of 2004. Within hours of the tsunami dozens of photos were available on the Web showing various affected places, and within days there were hundreds. As with the London bombings, there was no way to get photojournalists on the scene instantly, but here the problem was not just the speed of response but the spread of the damage, which affected thirteen countries. And as with the London bombings, the photos weren't used just for evidence; people began uploading photos of missing loved ones, and various weblogs began to syndicate these photos to aid in relocation. The most visited photo tagged "tsunami" is a picture of a little boy, age two at the time he went missing. The picture originally went up with contact information to aid in the search, but as time went on, it turned into an ongoing memorial; viewers posted hundreds of comments of support and prayers under the photo, and many commenters came back months later to check in and conversed with one another in the comments. When the boy's body was finally recovered and identified, months later, several people posted the sad news on Flickr, and the community that had formed around the photo posted expressions of grief and condolences for the family, then dissolved.

Photo sharing also helped provide the world with documentation of the 2006 military coup in Thailand. Immediately after the coup the military placed restrictions on reporting by the media, but it didn't (and probably couldn't) place similar restrictions on the whole populace. As a result, many of the earliest photos of tanks in front of Government House, the parliament building, came from individuals posting images from ordinary digital cameras, and they were discoverable by their tags (Bangkok, Thailand, Military, Coup). One of those users was Alisara Chirapongse, then a fashion-obsessed college student, who blogs under the name gnarlykitty, who posted the coup photos to her weblog, along with running commentary on the cause and immediate aftermath of the army overthrowing Thaksin Shinawatra, then prime minister. As the army announced that it wanted to take control of communications and ban public political speech, her posts took on a new urgency:

> One new little change that this law brought us is the whole new level of censorship. No political gathering, no discussing politics, and of course no voicing your opinions whatsoever about the whole mumbo jumbo coup. (Oops did I just do that?)

Alisara posted links to Wikipedia, the collaboratively produced encyclopedia, which was acting as a clearinghouse for breaking news of the coup (as is now usual). She also pointed her readers to a petition to restore freedom of speech and to a proposed demonstration, which she later attended and photographed.

Then, as the initial disorientation of the coup gave way to the new normal, Chirapongse went back to her life as a fashion-obsessed student. As she put it,

> This blog is my personal blog where I usually write things concerning my life and things I like. Since my

> life is lived here in Bangkok Thailand, it should come
> as no surprise to anyone that I sometime blog about
> it. So blogging about the Coup is merely blogging
> about something that's currently happening in
> my country.

The rest of that post was about a night she spent at a club, and the post after that was about how much she likes her new camera-phone. She wasn't a full-time journalist, she was a citizen with a camera and a weblog, but she had participated in a matter of global significance at exactly the time when the traditional media were being silenced.

The content in these examples is quite varied—the gentle ridiculousness of the Mermaid Parade and the awful seriousness of the London bombings; the man-made intervention of a military coup and the natural destruction of the tsunami. The common thread is the complexity of gathering the photos. The groups of photographers were all latent groups, which is to say groups that existed only *in potentia,* and too much effort would have been required to turn those latent groups into real ones by conventional means. The mermaid photos were too unimportant to be worth any institutional effort. The London bombing photos were taken by the people on the scene. The tsunami's destruction was spread out over tens of thousands of miles of coastland, and the uses of photos included finding missing persons, something outside the purview of typical newsgathering. During the Thai coup the military rulers were able to place restrictions on organized media, giving amateur photographers an advantage in providing views of tanks in the streets. In each of those cases the cost of coordinating the potential photographers would have defeated any institution wanting to put photos together quickly and make them available globally.

The task of aggregating and making photos available is nothing like, say, the task of putting a man on the moon. Prior to services like Flickr, what kept photo-sharing from happening wasn't the absolute difficulty but the relative difficulty. There is obviously some value to both photographers and viewers in having photos available, but in many cases that value never exceeded the threshold of cost created by the institutional dilemma. Flickr escaped those problems, not by increasing its managerial oversight over photographers but by abandoning any hope of such oversight in the first place, instead putting in place tools for the self-synchronization of otherwise latent groups.

## Making the Trains Run on Time

The structure of traditional managerial oversight is often illustrated by an "org chart," a diagram of the official organizational hierarchy. This chart is the simplest possible view of an organization's reporting structure. It is usually drawn as an inverted tree of boxes and arrows. The box at the top represents the head of the organization; the lines drawn downward from that box connect her to various officers and vice presidents through the layers of management, until, at the bottom, there are the rank and file, represented by boxes with lines connecting upward but not downward. The org chart diagrams both responsibility and channels of communication—when two boxes are connected on such a chart, the upper box is the boss; communication from the CEO flows down through the layers of management, while information from the workers flows up in the same way. Compared to the chaos of the market, the org chart draws clear and obvious lines of responsibility, and it is that very clarity that allows the firm to outperform a pure market for work.

The org chart is like institutional wallpaper—ubiquitous and not terribly dramatic. It's funny to think of it as a specific invention, but its existence and form owe quite a lot to the environment in which it was first widely used—railroad management in the 1800s. The pioneering managerial methods were meticulously documented by Alfred Chandler in his book *The Visible Hand*. The principal problem in running a railroad was arranging for eastbound and westbound trains to share the same track, because it was prohibitively expensive to lay more than one track for a particular line. By 1840 Western Railroad, a pioneer in building longer rail lines, had to deal with a dozen trains crossing in opposite directions every day. That situation created obvious safety risks, risks that were not long in moving from the theoretical to the real: on October 5, 1841, two passenger trains collided head on, with two fatalities and seventeen injured. This accident alarmed both the public and Congress and forced the railroads to rethink their management.

For the next fifteen years railroads invested in better oversight. As a result, their safety record improved, but their profitability decayed. A big firm like Western could haul more people and cargo to more places than could a smaller railroad, but the cost of managing the enterprise had risen much faster; Western was actually making less money per mile of track than its smaller competitors. David McCallum, a railroad superintendent for the New York & Erie Railroad, proposed both an explanation and a solution for this decayed profitability. As he put it in his Superintendent's Report of 1855:

> A Superintendent of a road fifty miles in length can give its business his personal attention, and may be constantly engaged in the direction of its details . . . any system however imperfect, may under such circumstances, prove comparatively successful.
>
> In the government of a road five hundred miles in length, a very different state exists. Any system which might be applicable to the business and extent of a short road, would be found entirely inadequate to the wants of a long one.

More is different: a small railroad could function with ad hoc management, because it had so few employees and so few passing trains, but as the scale rose, the management problems rose faster. This is where the institutional dilemma meets Birthday Paradox math: not only does managing resources take resources, but management challenges grow faster than organizational size.

McCallum's proposed solution to this dilemma included making a clear delineation of the responsibility for different segments of track. Central management would oversee regional divisions and supervise the trains passing through their region. McCallum introduced several formal innovations to New York & Erie: strong hierarchical oversight, including an explicitly divisional organization of the railroad with different superintendents responsible for different parts of the railroad. He diagrammed this form of organization with what may have been the first commercial org chart in history. This method was widely copied by other railroads, then by other kinds of firms.

In addition to revolutionizing management structure, McCallum wrote six principles for running a hierarchical organization. Most are what you'd expect (number one was ensuring a "proper division of responsibilities"), but number five is worth mentioning: his management system was designed to produce "such information, to be obtained through a system of daily reports and checks, that will not embarrass principal officers nor lessen their influence with their subordinates." If you have ever wondered why so much of what workers in large organizations know is shielded from the CEO and vice versa, wonder no longer: the idea of limiting communications, so that they flow only from one layer of the hierarchy to the next, was part of the very design of the system at the dawn of managerial culture.

## Post-Managerial Organization

When an organization takes on a task, the difficulty of coordinating everyone needs to be reined in somehow, and the larger the group, the more urgent the need. The standard, almost universal solution is to create a hierarchy and to slot individuals into that organization by role. In Coasean terms, McCallum's system lowered the transaction costs of running a railroad by increasing managerial structure. This approach greatly simplifies lines of responsibility and communication, making even very large organizations manageable. The individuals in such an organization have to agree to be managed, of course, which is usually achieved by paying them, and by making continued receipt of their pay contingent on their responsiveness to their manager's requests.

An organization will tend to grow only when the advantages that can be gotten from directing the work of additional employees are less than the transaction costs of managing them. Coase concentrated his analysis on businesses, but the problems of coordination costs apply to institutions of all sorts. The Catholic Church and the U.S. Army are as hierarchical as any for-profit business, and for many of the same reasons. The layers of structure between the pope and the priests, or between the president and the privates, is a product of the same forces as the layers between the general superintendent and a conductor on the New York & Erie. This hierarchical organization reduces transaction costs, but it doesn't eliminate them.

Imagine a company with fifteen hundred employees, where each manager is responsible for half a dozen people. The CEO has six vice presidents, who each direct the work of six supervisors, and so on. Such a company would have three layers of management between the boss and the workers. If you want to bring the workers closer to the boss, you will have to increase the number of workers that each manager is responsible for. This will reduce the number of layers but will also reduce average management time with each staff member (or force everyone to spend more hours per day communicating with one another). When an organization grows very large, it reaches the limit implicit in Coase's theory; at some point an institution simply cannot grow anymore and still remain functional, because the cost of managing the business will destroy any profit margin. You can think of this as a Coasean ceiling, the point above which standard institutional forms don't work well.

Coase's theory also tells us about the effects of small changes in transaction costs. When such costs fall moderately, we can expect to see two things. First, the largest firms increase in size. (Put another way, the upper limit of organizational size is inversely related to management costs.) Second, small companies become more effective, doing more business at lower cost than the same company does in a world of high transaction costs. These two effects describe the postwar industrial world well: Giant conglomerates like ITT in the 1970s and GE in recent years used their management acumen to get into a huge variety of businesses, simply because they were good at managing transaction costs. At the same time there has been an explosion of small- and medium-sized businesses, because such businesses were better able to discover and exploit new opportunities.

But what if transactions costs don't fall moderately? What if they collapse? This scenario is harder to predict from Coase's original work, and it used to be purely academic. Now it's not, because it's happening, or rather it has already happened, and we're starting to see the results.

Anyone who has worked in an organization with more than a dozen employees recognizes institutional costs. Anytime you are faced with too many meetings, too much paperwork, or too many layers of approval (shades of

McCallum), you are dealing with those costs. Until recently, such costs have been little more than the stuff of water-cooler grumbling—everyone complains about institutional overhead, without much hope of changing things. In that world (the world we lived in until recently), if you wanted to take on a task of any significance, managerial oversight was just one of the costs of doing business.

What happens to tasks that aren't worth the cost of managerial oversight? Until recently, the answer was "Those things don't happen." Because of transaction costs a long list of possible goods and services never became actual goods and services; things like aggregating amateur documentation of the London transit bombings were simply outside the realm of possibility. That collection now exists because people have always desired to share, and the obstacles that prevented sharing on a global scale are now gone. Think of these activities as lying under a Coasean floor; they are valuable to someone but too expensive to be taken on in any institutional way, because the basic and unsheddable costs of being an institution in the first place make those activities not worth pursuing.

Our basic human desires and talents for group effort are stymied by the complexities of group action at every turn. Coordination, organization, even communication in groups is hard and gets harder as the group grows. That difficulty means that whatever methods help coordinate group action will spread, no matter how inefficient they are, so long as they are better than nothing. Small groups have several methods for coordinating action, like calling each group member in turn or setting up a phone tree, but most of these methods don't work well even for dozens of people, much less for thousands. For large-scale activity, the methods that have worked best have been those pioneered by McCallum—hierarchical organization, managed in layers. The most common organizational structures we have today are simply the least bad fit for group action in an environment of high transaction costs.

Our new tools offer us ways of organizing group effort without resorting to McCallum's strategies. Flickr stands in a different kind of relationship to its photographers than a newspaper does. Where a newspaper is in the business of directing the work of photographers, Flickr is simply a platform; whatever coordination happens comes from the users and is projected onto the site. This is odd. We generally regard institutions as being capable of more things than uncoordinated groups are, precisely because they are able to direct their employees. Here, though, we have a situation where the loosely affiliated group can accomplish something more effectively than the institution can. Thanks to the introduction of user-generated labeling, the individual motivation of the photographers—devoid of financial reward—is now enough to bring vast collections of photos into being. These collections didn't just happen to be put together without an institution; that is the only way they could have been put together.

This is where Coasean logic gets strange. Small decreases in transaction costs make businesses more efficient, because the constraints of the institutional dilemma get less severe. Large decreases in transaction costs create activities that can't be taken on by businesses, or indeed by any institution, because no matter how cheap it becomes to perform a particular activity, there isn't enough payoff to support the cost incurred by being an institution in the first place. So long as the absolute cost of organizing a group is high, unmanaged groups will be limited to undertaking small efforts—a night out at the movies, a camping trip. Even something as simple as a potluck dinner typically requires some hosting institution. Now that it is possible to achieve large-scale coordination at low cost, a third category has emerged: serious, complex work, taken on without institutional direction. Loosely coordinated groups can now achieve things that were previously out of reach for any other organizational structure, because they lay under the Coasean floor.

The cost of all kinds of group activity—sharing, cooperation, and collective action—have fallen so far so fast that activities previously hidden beneath that floor are now coming to light. We didn't notice how many things were under that floor because, prior to the current era, the alternative to institutional action was usually no action. Social tools provide a third alternative: action by loosely structured groups, operating without managerial direction and outside the profit motive.

## From Sharing to Cooperation to Collective Action

For the last hundred years the big organizational question has been whether any given task was best taken on by the state, directing the effort in a planned way, or by businesses competing in a market. This debate was based on the universal and unspoken supposition that people couldn't simply self-assemble; the choice between markets and managed effort assumed that there was no third alternative. Now there is.

Our electronic networks are enabling novel forms of collective action, enabling the creation of collaborative groups that are larger and more distributed than at any other time in history. The scope of work that can be done by noninstitutional groups is a profound challenge to the status quo.

The collapse of transaction costs makes it easier for people to get together—so much easier, in fact, that it is changing the world. The lowering of these costs is the driving force underneath the current revolution and the common element to everything in this book. We're not used to thinking of "groupness" as a specific category—the differences between a college seminar and a labor union seem more salient than their similarities. It's hard to see how Evan Guttman's quest for the return of the mobile phone is the same kind of thing as the distributed documentation of the Indian Ocean tsunami. But like a chain of volcanoes all fed by the same pool of magma, the surface manifestations of group efforts seem quite separate, but the driving force of those eruptions is the same: the new ease of assembly. This change can be looked at as one long transition, albeit one with many manifestations, unfolding at different speeds in different contexts. The transition can be described in basic outline as the answer to two questions: Why has group action largely been limited to formal organizations? What is happening now to change that?

We now have communications tools—and increasingly, social patterns that make use of those tools—that are a better fit for our native desires and talents for group effort. Because we can now reach beneath the Coasean floor, we can have groups that operate with a birthday party's informality and a multinational's scope. What we are seeing, in the amateur coverage of the Thai coup and the tsunami documentation and the struggle over Ivanna's phone and countless other examples, is the beginning of a period of intense experimentation with these tools. The various results look quite different from one another, and as we get good at using the new tools, those results will diverge still further. New ease of assembly is causing a proliferation of effects, rather than a convergence, and these effects differ by how tightly the individuals are bound to one another in the various groups.

You can think of group undertaking as a kind of ladder of activities, activities that are enabled or improved by social tools. The rungs on the ladder, in order of difficulty, are sharing, cooperation, and collective action.

Sharing creates the fewest demands on the participants. Many sharing platforms, such as Flickr, operate in a largely take-it-or-leave-it fashion, which allows for the maximum freedom of the individual to participate while creating the fewest complications of group life. Though Flickr sets public sharing as the default, it also allows users to opt to show photos only to selected users, or to no one. Knowingly sharing your work with others is the simplest way to take advantage of the new social tools. (There are also ways of unknowingly sharing your work, as when Google reads the linking preferences of hundreds of millions of internet users. These users are helping create a communally available resource, as Flickr users are, but unlike Flickr, the people whose work Google is aggregating aren't actively choosing to make their contributions.)

Cooperation is the next rung on the ladder. Cooperating is harder than simply sharing, because it involves changing your behavior to synchronize with people who are changing their behavior to synchronize with you. Unlike sharing, where the group is mainly an aggregate of participants, cooperating creates group identity—you know who you are cooperating with. One simple form of cooperation, almost universal with social tools, is conversation; when people are in one another's company, even virtually, they like to talk. Sometimes the conversation is with words, as with e-mail, IM, or text messaging, and sometimes it is with other media: YouTube, the video sharing site, allows users to post new videos in response to videos they've seen on the site. Conversation creates more of a sense of community than sharing does, but it also introduces new problems. It is famously difficult to keep online conversations from devolving into either name-calling or blather, much less to keep them on topic. Some groups are perfectly happy with those effects (indeed, there are communities on the internet that revel in puerile or fatuous conversation), but for any group determined to maintain a set of communal standards some mechanism of enforcement must exist.

Collaborative production is a more involved form of cooperation, as it increases the tension between individual and group goals. The litmus test for collaborative production is simple: no one person can take credit for what gets created, and the project could not come into being without the participation of many. Structurally, the biggest difference between information sharing and collaborative production is that in collaborative production at least some collective decisions have to be made. The back-and-forth talking and editing that makes Wikipedia work results in a single page on a particular subject (albeit one that changes over time). Collaboration is not an absolute good—many tools work by *reducing* the amount of required coordination, as Flickr does in aggregating photos. Collaborative production can be valuable, but it is harder to get right than sharing, because anything that has to be negotiated about, like a Wikipedia article, takes more energy than things that can just be accreted, like a group of photos.

Collective action, the third rung, is the hardest kind of group effort, as it requires a group of people to commit themselves to undertaking a particular effort together, and to do so in a way that makes the decision of the group binding on the individual members. All group structures create dilemmas, but these dilemmas are hardest when it comes to collective action, because the cohesion of the group becomes critical to its success. Information sharing produces shared awareness among the participants, and collaborative production relies on shared creation, but collective action creates shared responsibility, by tying the user's identity to the identity of the group. In historical terms, a potluck dinner or a barn raising is collaborative production (the members work together to create something), while a union or a government engages in collective action, action that is undertaken in the name of the members meant to change something out in the world, often in opposition to other groups committed to different outcomes.

The commonest collective action problem is described as the "Tragedy of the Commons," biologist Garrett Hardin's phrase for situations wherein individuals have an incentive to damage the collective good. The Tragedy of the Commons is a simple pattern to explain, and once you understand it, you come to see it everywhere. The standard illustration of the problem uses sheep. Imagine you are one of a group of shepherds who graze their sheep on a commonly owned pasture. It's obviously in everyone's interest to keep the pasture healthy, which would require each of you to take care that your sheep don't overgraze. As long as everyone refuses to behave greedily, everyone benefits. There is just one problem with this system: "everyone" doesn't take your sheep to market. You do. Your incentive, as an individual shepherd, is to minimize the cost of raising the fattest possible sheep. Everyone benefits from you moderating your sheep's consumption of grass, but you would benefit more from free riding, which is to say letting them eat as much free grass as they possibly could.

Once you have this realization, you can still refrain from what would ultimately be a ruinous strategy, on the grounds that it would be bad for everyone else. Then another, even more awful thought strikes you: every other shepherd will have the same realization, and if even one of them decides to overgraze, all your good works will only end up subsidizing them. Seen in this light, the decision not to overgraze is provisional on everyone else making the same decision, which makes it very fragile indeed. The minute one of the other shepherds keeps his sheep out in the pasture an hour longer than necessary, the only power you have is to retaliate by doing the same. And this is the Tragedy of the Commons: while each person can agree that all would benefit from common restraint, the incentives of the individuals are arrayed against that outcome.

People who benefit from a resource while doing nothing in recompense are free riders. Societies have generally dealt with the problem of free riders in one of two ways. The first way is elimination of the commons, transferring ownership of parts of it to individuals, all of whom have an incentive to protect their own resources. If six shepherds each own one-sixth of the former commons, the overgrazing problem is a personal one, not a social one. If you overgraze your section, you will suffer the future consequences, while your neighbor will not. The second way is governance or, as Hardin puts it, "mutual coercion, mutually agreed upon." This solution prevents the individual actors from acting in their own interests rather than in the interests of the group. The Tragedy of the Commons is why taxes are never voluntary—people would opt out of paying for road maintenance if they thought their neighbors would pay for it. It's also why restaurants often add an automatic tip for large parties—when enough people are eating, everyone feels comfortable underfunding the group's tip, even if only unconsciously.

Collective action involves challenges of governance or, put another way, rules for losing. In any group that is determined to take collective action, different members of the group will express different opinions. Whenever a decision is taken on behalf of the group, at least some members won't get their way, and the bigger the group is, or the more decisions are made, the more often this will happen. For a group to take collective action, it must have some shared vision strong enough to bind the group together, despite periodic decisions that will inevitably displease at least some members. For this reason collective action is harder to arrange than information sharing or collaborative creation. In the current spread of social tools, real examples of collective action—where a group acts on behalf of, and with shared consequences for, all of its members—are still relatively rare.

The essential advantage created by new social tools has

been labeled "ridiculously easy group-forming" by the social scientist Seb Paquet. Our recent communications networks—the internet and mobile phones—are a platform for group-forming, and many of the tools built for those networks, from mailing lists to camera-phones, take that fact for granted and extend it in various ways. Ridiculously easy group-forming matters because the desire to be part of a group that shares, cooperates, or acts in concert is a basic human instinct that has always been constrained by transaction costs. Now that group-forming has gone from hard to ridiculously easy, we are seeing an explosion of experiments with new groups and new kinds of groups.

# CHAPTER 3

# EVERYONE IS A MEDIA OUTLET

*Our social tools remove older obstacles to public expression, and thus remove the bottlenecks that characterized mass media. The result is the mass amateurization of efforts previously reserved for media professionals.*

My uncle Howard was a small-town newspaperman, publishing the local paper for Richmond, Missouri (population 5,000). The paper, founded by my grandfather, was the family business, and ink ran in Howard's blood. I can still remember him fulminating about the rise of *USA Today;* he criticized it as "TV on paper" and held it up as further evidence of the dumbing down of American culture, but he also understood the challenge that *USA Today* presented, with its color printing and national distribution. The *Richmond Daily News* and *USA Today* were in the same business; even with the difference in scale and scope, Howard immediately got what *USA Today* was up to.

Despite my uncle's obsession, *USA Today* turned out to be

nothing like the threat that old-time newspaper people feared. It took some market share from other papers, but the effect wasn't catastrophic. What was catastrophic was a less visible but more significant change, already gathering steam when *USA Today* launched. The principal threat to the *Richmond Daily News*, and indeed to all newspapers small and large, was not competition from other newspapers but radical changes in the overall ecosystem of information. The idea that someone might build four-color presses that ran around the clock was easy to grasp. The idea that the transmission of news via paper might become a bad idea, that all those huge, noisy printing presses might be like steam engines in the age of internal combustion, was almost impossible to grasp. Howard could imagine someone doing what he did, but better. He couldn't imagine someone making what he did obsolete.

Many people in the newspaper business, the same people who worried about the effects of competition like *USA Today*, missed the significance of the internet. For people with a professional outlook, it's hard to understand how something that isn't professionally produced could affect them—not only is the internet not a newspaper, it isn't a business, or even an institution. There was a kind of narcissistic bias in the profession; the only threats they tended to take seriously were from other professional media outlets, whether newspapers, TV, or radio stations. This bias had them defending against the wrong thing when the amateurs began producing material on their own. Even as web sites like eBay and Craigslist were siphoning off the ad revenues that keep newspapers viable—job listings, classified ads, real estate—and weblogs were letting people like gnarlykitty publish to the world for free, the executives of the world's newspapers were slow to understand the change, and even slower to react. How could this happen? How could the newspaper industry miss such an obvious and grave challenge to their business? The answer is the flip side of Howard's obsession with *USA Today* and has to do with the nature of professional self-definition (and occasional self-delusion).

A profession exists to solve a hard problem, one that requires some sort of specialization. Driving a race car requires special training—race car drivers are professionals. Driving an ordinary car, though, doesn't require the driver to belong to a particular profession, because it's easy enough that most adults can do it with a modicum of training. Most professions exist because there is a scarce resource that requires ongoing management: librarians are responsible for organizing books on the shelves, newspaper executives are responsible for deciding what goes on the front page. In these cases, the scarcity of the resource itself creates the need for a professional class— there are few libraries but many patrons, there are few channels but many viewers. In these cases professionals become gatekeepers, simultaneously providing and controlling access to information, entertainment, communication, or other ephemeral goods.

To label something a profession means to define the ways in which it is more than just a job. In the case of newspapers, professional behavior is guided both by the commercial imperative and by an additional set of norms about what newspapers are, how they should be staffed and run, what constitutes good journalism, and so forth. These norms are enforced not by the customers but by other professionals in the same business. The key to any profession is the relations of its members to one another. In a profession, members are only partly guided by service to the public. As the UCLA sociologist James Q. Wilson put it in his magisterial *Bureaucracy*, "A professional is someone who receives important occupational rewards from a reference group whose membership is limited to people who have undergone specialized formal education and have accepted a group-defined code of proper conduct." That's a mouthful, but the two key ideas apply to newspaper publishers (as well as to journalists, lawyers, and accountants): a professional learns things in a way that differentiates her from most of the populace, and she pays as much or more attention to the judgment of her peers as to the judgment of her customers when figuring out how to do her job.

A profession becomes, for its members, a way of understanding their world. Professionals see the world through a lens created by other members of their profession; for journalists, the rewards of a Pulitzer Prize are largely about recognition from other journalists.

Much of the time the internal consistency of professional judgment is a good thing—not only do we want high standards of education and competence, we want those standards created and enforced by other members of the same profession, a structure that is almost the definition of professionalism. Sometimes, though, the professional outlook can become a disadvantage, preventing the very people who have the most at stake—the professionals themselves—from understanding major changes to the structure of their profession. In particular, when a profession has been created as a result of some scarcity, as with librarians or television programmers, the professionals are often the last ones to see it when that scarcity goes away. It is easier to understand that you face competition than obsolescence.

In any profession, particularly one that has existed long enough that no one can remember a time when it didn't exist, members have a tendency to equate provisional solutions to particular problems with deep truths about the world. This is true of newspapers today and of the media generally. The media industries have suffered first and most from the recent collapse in communications costs. It used to be hard to move words, images, and sounds from creator to consumer, and most media businesses involve expensive and complex management of that pipeline problem, whether running a printing press or a record label. In return for helping overcome these problems, media businesses got to exert considerable control over the media and extract considerable revenues from the public. The commercial viability of most media businesses involves providing those solutions, so preservation of the original problems became an economic imperative. Now, though, the problems of production, reproduction, and distribution are much less serious. As a consequence, control over the media is less completely in the hands of the professionals.

As new capabilities go, unlimited perfect copyability is a lulu, and that capability now exists in the hands of everyone who owns a computer. Digital means of distributing words and images have robbed newspapers of the coherence they formerly had, revealing the physical object of the newspaper as a merely provisional solution; now every article is its own section. The permanently important question is how society will be informed of the news of the day. The newspaper used to be a pretty good answer to that question, but like all such answers, it was dependent on what other solutions were available. Television and radio obviously changed the landscape in which the newspaper operated, but even then printed news had a monopoly on the written word—until the Web came along. The Web didn't introduce a new competitor into the old ecosystem, as *USA Today* had done. The Web created a new ecosystem.

We've long regarded the newspaper as a sensible object because it has been such a stable one, but there isn't any logical connection among its many elements: stories from Iraq, box scores from the baseball game, and ads for everything from shoes to real estate all exist side by side in an idiosyncratic bundle. What holds a newspaper together is primarily the cost of paper, ink, and distribution; a newspaper is whatever group of printed items a publisher can bundle together and deliver profitably. The corollary is also true: what doesn't go into a newspaper is whatever is too expensive to print and deliver. The old bargain of the newspaper—world news lumped in with horoscopes and ads from the pizza parlor—has now ended. The future presented by the internet is the mass amateurization of publishing and a switch from "Why publish this?" to "Why not?"

The two basic organizational imperatives—acquire resources, and use them to pursue some goal or agenda—saddle every organization with the institutional dilemma, whether its goal is saving souls or selling soap. The question that mass amateurization poses to traditional media is "What happens when the costs of reproduction and distribution go away? What happens when there's nothing unique about publishing any-more, because users can do it for themselves?" We are now starting to see that question being answered.

## Weblogs and Mass Amateurization

Shortly after his reelection in 2002 Trent Lott, the senior senator from Mississippi and then majority leader, gave a speech at Strom Thurmond's hundredth birthday party. Thurmond, a Republican senator from South Carolina, had recently retired after a long political career, which had included a 1948 run for president on an overtly segregationist platform. At Thurmond's hundredth birthday party Lott remembered and praised Thurmond's presidential campaign of fifty years earlier and recalled Mississippi's support for it: "I want to say this about my state: When Strom Thurmond ran for president, we voted for him. We're proud of it. And if the rest of the country had followed our lead, we wouldn't have had all these problems over all these years, either." Two weeks later, having been rebuked by President Bush and by politicians and the press on both the right and the left for his comment, Lott announced that he would not seek to remain majority leader in the new Congress.

This would have been a classic story of negative press cov-erage altering a political career—except that the press didn't actually cover the story, at least not at first. Indeed, the press almost completely missed the story. This isn't to say that they intentionally ignored it or even actively suppressed it; several reporters from national news media heard Lott speak, but his remark simply didn't fit the standard template of news. Because Thurmond's birthday was covered as a salutary event instead of as a political one, the actual contents of the evening were judged in advance to be relatively unimportant. A related assumption is that a story that is not important one day also isn't important the next, unless something has changed. Thurmond's birthday party happened on a Thursday night, and the press gave Lott's remarks very little coverage on Friday. Not having written about it on Friday in turn became a reason not to write about it on Saturday, because if there was no story on Friday, there was even less of one on Saturday.

William O'Keefe of *The Washington Post*, one of the few reporters to think Lott's comment was important, explains the dilemma this way: "[T]here had to be a reaction" that the network could air alongside Lott's remarks, and "we had no on-camera reaction" available the evening of the party, when the news was still fresh. By the following night, he adds, "you're dealing with the news cycle: twenty-four hours later—that's old news." Like a delayed note to a friend, the initial lack of response would have meant, in any later version, having to apologize for not having written sooner.

Given this self-suppression—old stories are never revisited without a new angle—what kept the story alive was not the press but liberal and conservative bloggers, for whom fond memories of segregation were beyond the pale, birthday felicitations or no, and who had no operative sense of news cycles. The weekend after Lott's remarks, weblogs with millions of readers didn't just report his comments, they began to editorialize. The editorializers included some well-read conservatives such as Glenn Reynolds of the Instapundit blog, who wrote, "But to say, as Lott did, that the country would be better off if Thurmond had won in 1948 is, well, it's proof that Lott shouldn't be majority leader for the Republicans, to begin with. And that's just to begin with. It's a sentiment as evil and loony as wishing that Gus Hall [a perennial Communist candidate for president] had been elected."

Even more damaging to Lott, others began to dig deeper. After the story broke, Ed Sebesta, who maintains a database of materials related to nostalgia for the U.S. Confederacy, contacted bloggers with information on Lott, including an interview from the early 1980s in *Southern Partisan,* a neo-Confederate magazine. The simple birthday party story began looking like part of a decades-long pattern of saying one thing to the general public and another thing to his supporters.

Like the story of Ivanna's lost phone (in Chapter 1), the story of Sebesta's database involves a link between individual effort and group attention. Just as Evan Guttman benefited from the expert knowledge of his readers, the bloggers posting about Lott benefited from Sebesta's deep knowledge of America's racist past, particularly of Lott's history of praise for same. Especially important, the bloggers didn't have to find Sebesta—he found them. Prior to our current generation of coordinating tools, a part-time politics junkie like Sebesta and amateur commentators like the bloggers would have had a hard time even discovering that they had mutual interests, much less being able to do anything with that information. Now, however, the cost of finding like-minded people has been lowered and, more important, deprofessionalized.

Because the weblogs kept the story alive, especially among libertarian Republicans, Lott eventually decided to react. The fateful moment came five days after the speech, when he issued a halfhearted apology for his earlier remark, characterizing it as a "poor choice of words." The statement was clearly meant to put the matter behind him, but Lott had not reckoned with the changed dynamics of press coverage. Once Lott apologized, news outlets could cover the apology as the news, while quoting the original speech as background. Only three mainstream news outlets had covered the original comment, but a dozen covered the apology the day it happened, and twenty-one covered it the day after. The traditional news cycle simply didn't apply in this situation; the story had suddenly been transformed from "not worth covering" to "breaking news."

Until recently, "the news" has meant two different things—events that are newsworthy, and events covered by the press. In that environment what identified something as news was professional judgment. The position of the news outlets (the very phrase attests to the scarcity of institutions that were able to publish information) was like that of the apocryphal umpire who says, "Some pitches are balls and some are strikes, but they ain't nothin' till I call 'em." There has always been grumbling about this system, on the grounds that some of the things the press was covering were not newsworthy (politicians at ribbon cuttings) and that newsworthy stories weren't being covered or covered enough (insert your pet issue here). Despite the grumbling, however, the basic link between newsworthiness and publication held, because there did not seem to be an alternative. What the Lott story showed us was that the link is now broken. From now on news can break into public consciousness without the traditional press weighing in. Indeed, the news media can end up covering the story *because* something has broken into public consciousness via other means.

There are several reasons for this change. The professional structuring of worldview, as exemplified by the decisions to treat Lott's remarks as a birthday party story, did not extend to the loosely coordinated amateurs publishing on their own. The decision not to cover Trent Lott's praise for a racist political campaign demonstrates a potential uniformity in the press outlook. In a world where a dozen editors, all belonging to the same professional class, can decide whether to run or kill a national story, information that might be of interest to the general public may not be published, not because of a conspiracy but because the editors have a professional bias that is aligned by the similar challenges they face and by the similar tools they use to approach those challenges. The mass amateurization of publishing undoes the limitations inherent in having a small number of traditional press outlets.

As they surveyed the growing amount of self-published content on the internet, many media companies correctly understood that the trustworthiness of each outlet was lower than that of established outlets like *The New York Times*. But what they failed to understand was that the effortlessness of publishing means that there are many more outlets. The same idea, published in dozens or hundreds of places, can have an amplifying effect that outweighs the verdict from the smaller number of professional outlets. (This is not to say that mere repetition makes an idea correct; amateur publishing relies on corrective argument even more than traditional media do.) The change isn't a shift from one kind of news institution to another, but rather in the definition of news: from news as an institutional prerogative to news as part of a communications ecosystem, occupied by a mix of formal organizations, informal collectives, and individuals.

It's tempting to regard the bloggers writing about Trent Lott or the people taking pictures of the Indian Ocean tsunami as a new crop of journalists. The label has an obvious conceptual appeal. The problem, however, is that mass professionalization is an <u>oxymoron</u>, since a professional class implies a specialized function, minimum tests for competence, and a minority of members. None of those conditions exist with political weblogs, photo sharing, or a host of other self-publishing tools. The individual weblogs are not merely alternate sites of publishing; they are alternatives to publishing itself, in the sense of publishers as a minority and professional class. In the same way you do not have to be a professional driver to drive, you no longer have to be a professional publisher to publish. Mass amateurization is a result of the radical spread of expressive capabilities, and the most obvious precedent is the one that gave birth to the modern world: the spread of the printing press five centuries ago.

## In Praise of Scribes

Consider the position of a scribe in the early 1400s. The ability to write, one of the crowning achievements of human inventiveness, was difficult to attain and, as a result, rare. Only a tiny fraction of the populace could actually write, and the wisdom of the ages was encoded on fragile and decaying manuscripts.

In this environment a small band of scribes performed the essential service of refreshing cultural memory. By hand-copying new editions of existing manuscripts, they performed a task that could be performed no other way. The scribe was the only bulwark against great intellectual loss. His function was indispensable, and his skills were irreplaceable.

Now consider the position of the scribe at the end of the 1400s. Johannes Gutenberg's invention of movable type in the middle of the century had created a sudden and massive reduction in the difficulty of reproducing a written work. For the first time in history a copy of a book could be created faster than it could be read. A scribe, someone who has given his life over to literacy as a cardinal virtue, would be conflicted about the meaning of movable type. After all, if books are good, then surely more books are better. But at the same time the very scarcity of literacy was what gave scribal effort its primacy, and the scribal way of life was based on this scarcity. Once the scribe's skills were eminently replaceable, his function—making copies of books—was better accomplished by ignoring tradition than by embracing it.

Two things are true about the remaking of the European intellectual landscape during the Protestant Reformation: first, it was not caused by the invention of movable type, and second, it was possible only after the invention of movable type, which aided the rapid dissemination of Martin Luther's complaints about the Catholic Church (the 95 *Theses*) and the spread of Bibles printed in local languages, among its other effects. Holding those two thoughts in your head at the same time is essential to understanding any social change driven by a new technological capability. Because social effects lag behind technological ones by decades, real revolutions don't involve an orderly transition from point A to point B. Rather, they go from A through a long period of chaos and only then reach B. In that chaotic period, the old systems get broken long before new ones become stable. In the late 1400s scribes existed side by side with publishers but no longer performed an irreplaceable service. Despite the replacement of their core function, however, the scribes' sense of themselves as essential remained undiminished.

In 1492, almost half a century after movable type appeared, Johannes Trithemius, the Abbot of Sponheim, was moved to launch an impassioned defense of the scribal tradition, *De Laude Scriptorum* (literally "in praise of scribes"). In this work he laid out the values and virtues of the scribal tradition: "The devout monk enjoys four particular benefits from writing: the time that is precious is profitably spent; his understanding is enlightened as he writes; his heart within is kindled to devotion; and after this life he is rewarded with a unique prize." Note how completely the benefits of the scribal tradition are presented as ones enjoyed by scribes rather than by society.

The Abbot's position would have been mere reactionary cant ("We must preserve the old order at any cost") but for one detail. If, in the year 1492, you'd written a treatise you wanted widely disseminated, what would you do? You'd have it printed, of course, which was exactly what the Abbot did. *De Laude Scriptorum* was not itself copied by scribes; it was set in movable type, in order to get a lot of copies out cheaply and quickly— something for which scribes were utterly inadequate. The content of the Abbot's book praised the scribes, while its printed form damned them; the medium undermined the message.

There is an instructive hypocrisy here. A professional often becomes a gatekeeper, by providing a necessary or desirable social function but also by controlling that function. Sometimes this gatekeeping is explicitly enforced (only judges can sentence someone to jail, only doctors can perform surgery) but sometimes it is embedded in technology, as with scribes, who had mastered the technology of writing. Considerable effort must be expended toward maintaining the discipline and structure of the profession. Scribes existed to increase the spread of the written word, but when a better, nonscribal way of accomplishing the same task came along, the Abbot of Sponheim stepped in to argue that preserving the scribes' way of life was more important than fulfilling their mission by nonscribal means.

Professional self-conception and self-defense, so valuable in ordinary times, become a disadvantage in revolutionary ones, because professionals are always concerned with threats to the profession. In most cases, those threats are also threats to society; we do not want to see a relaxing of standards for becoming a surgeon or a pilot. But in some cases the change that threatens the profession benefits society, as did the spread of the printing press; even in these situations the professionals can be relied on to care more about self-defense than about progress. What was once a service has become a bottleneck. Most organizations believe they have much more freedom of action and much more ability to shape their future than they actually do, and evidence that the ecosystem is changing in ways they can't control usually creates considerable anxiety, even if the change is good for society as a whole.

## Mass Amateurization Breaks
## Professional Categories

Today the profession of scribe seems impossibly quaint, but the habit of tying professional categories to mechanical processes is alive and well. The definition of journalist, seemingly a robust and stable profession, turns out to be tied to particular forms of production as well.

In 2006 Judith Miller, then a reporter for *The New York Times*, was jailed for eighty-five days for refusing to reveal her sources in an ongoing federal investigation, becoming a cause célèbre for reporters in the United States. She eventually relented, after those sources released her from any expectation of confidentiality, and was freed, but by that time her incarceration had created a great deal of unease about the fate of journalistic privilege—the right of journalists to grant promises of confidentiality in order to convince potential sources to cooperate. Though some sort of shield law for journalists exists in forty-nine of fifty states, federal law has no equivalent. Seeing the risk of federal incarceration without such protection, several members of Congress introduced bills to create a federal shield law. Surprisingly, though, what seemed like a simple technicality—pass the same kind of law at the federal level as existed in most of the states—turned out to be not merely complex but potentially impossible, and the difficulties stemmed from a simple question: who, exactly, should enjoy journalistic privilege?

The tautological answer is that journalists should enjoy such privileges, but who are journalists? One view defines "journalist," in the words of the *Oxford English Dictionary,* as "a person who writes for newspapers or magazines or prepares news to be broadcast on radio or television." This is an odd definition, as it provides less a description of journalism than a litmus test of employment. In this version, journalists aren't journalists unless they work for publishers, and publishers aren't publishers unless they own the means of production. This definition has worked for decades, because the ties among journalists, publishers, and the means of production were strong. So long as publishing was expensive, publishers would be rare. So long as publishers were rare, it would be easy to list them and thus to identify journalists as their employees. This definition, oblique as it is, served to provide the legal balance we want from journalistic privilege—we have a professional class of truth-tellers who are given certain latitude to avoid co-operating with the law. We didn't have to worry, in defining those privileges, that they would somehow become general, because it wasn't like just anyone could become a publisher.

And now it is like that. It's exactly like that. To a first approximation, anyone in the developed world can publish anything anytime, and the instant it is published, it is globally available and readily findable. If anyone can be a publisher, then anyone can be a journalist. And if anyone can be a journalist, then journalistic privilege suddenly becomes a loophole too large to be borne by society. Journalistic privilege has to be applied to a minority of people, in order to preserve the law's ability to uncover and prosecute wrongdoing while allowing a safety valve for investigative reporting. Imagine, in a world where any blogger could claim protection, trying to compel someone to testify about their friend's shady business: "Oh,

I can't testify about that. I've been blogging about it, so what he told me is confidential."

We can't just exclude bloggers either. Many well-read bloggers are journalists, like the war reporter Kevin Sites, who was fired from CNN for blogging, then went to blog on his own; or Rebecca Mackinnon, who was formerly at CNN and went on to cofound Global Voices, dedicated to spreading blogging throughout the world; or Dan Gillmor, a journalist at the *San Jose Mercury News* who blogged both during and after his tenure; and so on. It's tempting to grandfather these bloggers as journalists, since they were journalists before they were blogging, but that would essentially be to ignore the weblog as a form, since a journalist would have to be anointed by some older form of media. This idea preserves what is most wrong with the original definition, namely that the definition of journalist is not internally consistent but rather is tied to ownership of communications machinery. Such a definition would exclude Ethan Zuckerman, a cofounder of Global Voices with Mackinnon; it's hard to imagine any sensible definition of journalist that would include her and exclude him, but it's also hard to imagine any definition that includes him without opening the door to including tens of millions of bloggers, too large a group to be acceptable. It would exclude Xeni Jardin, one of the contributors to the well-trafficked weblog Boing Boing who, as a result of her blogging, has gotten a spot on NPR. Did she become a journalist after NPR anointed her? Did her blogging for Boing Boing become journalism afterward? What about the posts from before—did they retroactively become the work of a journalist? And so on.

The simple answer is that there is no simple answer.

Journalistic privilege is based on the previous scarcity of publishing. When it was easy to recognize who the publisher was, it was easy to figure out who the journalists were. We could regard them as a professional (and therefore minority) category. Now that scarcity is gone. Facing the new abundance of publishing options, we could just keep adding to the list of possible outlets to which journalism is tied—newspapers and television, and now blogging and video blogging and podcasting and so on. But the latter items on the list are different because they have no built-in scarcity. Anyone can be a publisher (and frequently is). There is never going to be a moment when we as a society ask ourselves, "Do we want this? Do we want the changes that the new flood of production and access and spread of information is going to bring about?" It has already happened; in many ways, the rise of group-forming networks is best viewed not as an invention but as an event, a thing that has happened in the world that can't be undone. As with the printing press, the loss of professional control will be bad for many of society's core institutions, but it's happening anyway. The comparison with the printing press doesn't suggest that we are entering a bright new future—for a hundred years after it started, the printing press broke more things than it fixed, plunging Europe into a period of intellectual and political chaos that ended only in the 1600s.

This issue became more than academic with the arrest of Josh Wolf, a video blogger who refused to hand over video of a 2005 demonstration he observed in San Francisco. He served 226 days in prison, far longer than Judith Miller, before being released. In one of his first posts after regaining his freedom, he said, "The question that needs to be asked is not 'Is Josh Wolf

a journalist?' but 'Should journalists deserve the same protec-
tions in federal court as those afforded them in state courts?"
This isn't right, though, because making the assumption that
Wolf is a journalist in any uncomplicated way breaks the social
expectations around journalism in the first place. The question
that needs to be asked is, "Now that there is no limit to those
who can commit acts of journalism, how should we alter journalistic privilege to fit that new reality?" The admission of Wolf
into the category of journalist breaks the older version of that
category, giving the question "Who is a journalist?" a new complexity.

The pattern is easy to see with journalists, but it isn't restricted to them. Who is a professional photographer? Like
"journalist," that category seems at first to be coherent and
internally cohesive, but it turns out to be tied to scarcity as
well. The amateurization of the photographers' profession
began with the spread of cheap cameras generally, but it really
took off with digital photos and online photo hosting sites.
The threat to professional photographers came from a change
not just in the way photographs were created but in the way
they were distributed. In contrast to the situation a few years
ago, taking and publishing photographs doesn't even require
the purchase of a camera (mobile phones already sport
surprisingly high-quality digital cameras), and it certainly
doesn't require access either to a darkroom or to a special
publishing outlet. With a mobile phone and a photo-sharing
service, people are now taking photographs that are being
seen by thousands and, in rare cases, by millions of people, all
without any money changing hands.

The twin effects are an increase in good amateur photographs and a threat to the market for professionals. Jeff Howe, author of the forthcoming *Crowdsourcing*, describes iStockPhoto.com, a Web-based clearinghouse for photographers to offer their work for use in advertising and promotional materials (a practice called stock photography). Prior to services like iStockPhoto, amateurs had no outlet for selling their photos, no matter what the quality, leaving the market to professionals. Because one of the services provided by professionals was the simple availability and findability of their photos relative to the amateurs, they commanded a premium for each photo sold. How high was that premium? When a project director at the National Health Museum wanted pictures of flu sufferers, Howe notes, the price from a professional photographer was over $100 (after a discount) per photo, while the price from iStockPhoto was one dollar, less than one percent of the professional's price. Much of the price for professional stock photos came from the difficulty of finding the right photo rather than from the difference in quality between photos taken by professionals and amateurs. The success of iStockPhoto suggests that the old division of amateur and professional is only a gradient rather than a gap and that it can be calculated photo by photo. If an amateur has taken only one good photo in his life, but you can find it, why not use it? As with the profession of journalist, iStockPhoto shows that the seemingly consistent profession of photographer is based on criteria that are external to the profession itself. The only real arbiter of professionalism in photography today is the taxman; in the United States, the IRS defines a professional photographer as someone who makes more than $5,000 a year selling his or her photos.

New communications capabilities are also changing social

definitions that are not tied to professions. Consider what happened to Sherron Watkins, an accountant at the failed energy firm Enron. In 2001 Watkins wrote an e-mail to a handful of executives at Enron and their accounting firm entitled, "The smoking gun you can't extinguish," wherein she detailed the dangerous practices Enron was using to hide its true revenues and costs. As Watkins put it presciently, "I am incredibly nervous that we will implode in a wave of accounting scandals," which is exactly what happened the following year. Watkins was widely described as a whistle-blower, even though her e-mail was addressed to only a handful people at Enron and at the accounting firm Arthur Andersen. Different from any previous definition of whistle-blower, all Watkins did was write a particularly damning interoffice memo; she didn't leak anything to the press. What the application of the whistle-blower label signals is that in an age of infinite perfect copyability to many people at once, the very act of writing and sending an e-mail can be a kind of publishing, because once an e-mail is sent, it is almost impossible to destroy all the copies, and anyone who has a copy can broadcast it to the world at will, and with ease. Now, and presumably from now on, the act of creating and circulating evidence of wrongdoing to more than a few people, even if they all work together, will be seen as a delayed but public act.

The pattern here is simple—what seems like a fixed and abiding category like "journalist" turns out to be tied to an accidental scarcity created by the expense of publishing apparatus. Sometimes this scarcity is decades old (as with photographers) or even centuries old (as with journalists), but that doesn't stop it from being accidental, and when that scarcity gets undone, the seemingly stable categories turn out to be unsupportable. This is not to say that professional journalists and photographers do not exist—no one is likely to mistake Bob Woodward or Annie Liebowitz for an amateur—but it does mean that the primary distinction between the two groups is gone. What once was a chasm has now become a mere slope.

Publishing used to require access to a printing press, and as a result the act of publishing something was limited to a tiny fraction of the population, and reaching a population outside a geographically limited area was even more restricted. Now, once a user connects to the internet, he has access to a platform that is at once global and free. It isn't just that our communications tools are cheaper; they are also better. In particular, they are more favorable to innovative uses, because they are considerably more flexible than our old ones. Radio, television, and traditional phones all rely on a handful of commercial firms owning expensive hardware connected to cheap consumer devices that aren't capable of very much. The new model assumes that the devices themselves are smart; this in turn means that one may propose and explore new models of communication and coordination without needing to get anyone's permission first (to the horror of many traditional media firms). As Scott Bradner, a former trustee of the Internet Society, puts it, "The internet means you don't have to convince anyone else that something is a good idea before trying it."

An individual with a camera or a keyboard is now a nonprofit of one, and self-publishing is now the normal case. This spread has been all the more remarkable because this technological story is not like the story of the automobile, where an invention moved from high cost to low cost, so that it went from being a luxury to being a commonplace possession. Rather, this technological story is like literacy, wherein a particular capability moves from a group of professionals to become embedded within society itself, ubiquitously, available to a majority of citizens.

When reproduction, distribution, and categorization were all difficult, as they were for the last five hundred years, we needed professionals to undertake those jobs, and we properly venerated those people for the service they performed. Now those tasks are simpler, and the earlier roles have in many cases become optional, and are sometimes obstacles to direct access, often putting the providers of the older service at odds with their erstwhile patrons. An amusing example occurred in 2005, when a French bus company, Transports Schiocchet Excursions (TSE), sued several French cleaning women who had previously used TSE for transport to their jobs in Luxembourg. The women's crime? Carpooling. TSE asked that the women be fined and that their cars be confiscated, on the grounds that the service the women had arranged to provide for themselves—transportation—should be provided only by commercial services such as TSE. (The case was thrown out in a lower court; it is pending on appeal.)

Though this incident seems like an unusual lapse in business judgment, this strategy—suing former customers for organizing themselves—is precisely the one being pursued by the music and movie industries today. Those industries used to perform a service by distributing music and moving images, but laypeople can now move music and video easily, in myriad ways that are both cheaper and more flexible than those mastered and owned by existing commercial firms, like selling CDs and DVDs in stores. Faced with these radical new efficiencies, those very firms are working to make moving movies and music harder, in order to stay in business—precisely the outcome that the bus company (and the Abbott) was arguing for.

In a world where publishing is effortless, the decision to publish something isn't terribly momentous. Just as movable type raised the value of being able to read and write even as it destroyed the scribal tradition, globally free publishing is making public speech and action more valuable, even as its absolute abundance diminishes the specialness of professional publishing. For a generation that is growing up without the scarcity that made publishing such a serious-minded pursuit, the written word has no special value in and of itself. Adam Smith, in *The Wealth of Nations*, pointed out that although water is far more important than diamonds to human life, diamonds are far more expensive, because they are rare. The entire basis on which the scribes earned their keep vanished not when reading and writing vanished but when reading and writing became ubiquitous. If everyone can do something, it is no longer rare enough to pay for, even if it is vital.

The spread of literacy after the invention of movable type ensured not the success of the scribal profession but its end. Instead of mass professionalization, the spread of literacy was a process of mass amateurization. The term "scribe" didn't get extended to everyone who could read and write. Instead, it simply disappeared, as it no longer denoted a professional class. The profession of calligrapher now survives as a purely decorative art; we make a distinction between the general ability to write and the professional ability to write in a calligraphic hand, just as we do between the general ability to drive and the professional ability to drive a race car. This is what is happening today, not just to newspapers or to media in general but to the global society.

# CHAPTER 4

# PUBLISH, THEN FILTER

*The media landscape is transformed, because personal communication and publishing, previously separate functions, now shade into one another. One result is to break the older pattern of professional filtering of the good from the mediocre before publication; now such filtering is increasingly social, and happens after the fact.*

Here, on a random Tuesday afternoon in May, is some of what is on offer from the world's mass of amateurs.

At LiveJournal, Kelly says:

> yesterdayyyyy, after the storm of the freaking century, i went to the mall with deanna, dixon and chris. we ran into everyone in the world there, got food, and eventually picked out clothes for dixon. found katie and ryan and forced katie to come back to my house with me and dixon. then deanna came a little after, then jimmy pezz, and then lynn. good times,

> good times. today, i woke up to my dog barking like
> a maniac and someone knocking on my window. i
> was so freaked out, but then jackii told me it was
> jack so i was just like whatever and went back to
> sleep. i have no idea what im doing today but par-
> tyyy tonighttt

At YouTube, texasgirly1979's twenty-six-second video of a pit bull nudging some baby chicks with his nose has been viewed 1,173,489 times.

At MySpace, a user going by Loyonon posts a message on Julie's page:

> Julieeeeeeeeeee I can't believe I missed you last
> night!!! Trac talked to you and said you were
> TRASHED off your ASS! Damn, I missed it. lol
> ["laughing out loud"]

At Flickr, user Frecklescorp has uploaded a picture of a woman at a fancy dress party, playing a ukulele.

At Xanga, user Angel_An_Of_Lips says:

> Hey every1 srry i havent been on a while i have been
> caught up in a lot of things like softball and volleyball
> my new dog and im goin to Tenn. on thursday so i
> wont be on here for bout a week but i promise i will
> get on and show pic. and michigan was so funnnn~!
> welp we got a jack russel terrier and this is wut it
> looks like!!. . . . . . . . . isnt he sooo cute. . . . i no!!!

welp thats all i got to say oh oh yah i got my hair cut
it is in my pic. cool uhh . . . ~!

And that, of course, is a drop in the bucket. Surveying this vast collection of personal postings, in-joke photographs, and poorly shot video, it's easy to conclude that, while the old world of scarcity may had some disadvantages, it spared us the worst of amateur production. Surely it is as bad to gorge on junk as to starve?

The catchall label for this material is "user-generated content." That phrase, though, is something of a misnomer. When you create a document on your computer, your document fits some generic version of the phrase, but that isn't really what user-generated content refers to. Similarly, when Stephen King composes a novel on his computer, that isn't user-generated content either, even though Mr. King is a user of software just as surely as anyone else. User-generated content isn't just the output of ordinary people with access to creative tools like word processors and drawing programs; it requires access to *re*-creative tools as well, tools like Flickr and Wikipedia and weblogs that provide those same people with the ability to distribute their creations to others. This is why the file on your computer doesn't count as user-generated content—it doesn't find its way to an audience. It is also why Mr. King's novel-in-progress doesn't count—he is paid to get an audience. User-generated content is a group phenomenon, and an amateur one. When people talk about user-generated content, they are describing the ways that users create and share media with one another, with no professionals anywhere in sight. Seen this way, the idea of user-generated content is actually not just a personal theory of creative capabilities but a social theory of media relations.

MySpace, the wildly successful social networking site, has tens of millions of users. We know this because the management of MySpace (and of its parent company, News Corp) tells the public how many users they have at every opportunity. But most users don't experience MySpace at the scale of tens of millions. Most users interact with only a few others—the median number of friends on MySpace is two, while the average number of "friends" is fifty-five. (That latter figure is in quotes because the average is skewed upward by individuals who list themselves as "friends" of popular bands or of the site's founder, Tom.) Even this average of fifty-five friends, skewed upward as it is, demonstrates the imbalance: the site has had more than a hundred million accounts created, but most people link to a few dozen others at most. No one (except News Corp) can easily address the site's assembled millions; most conversation goes on in much smaller groups, albeit interconnected ones. This pattern is general to services that rely on social networking, like Facebook, LiveJournal, and Xanga. It is even true of the weblog world in general—dozens of weblogs have an audience of a million or more, and millions have an audience of a dozen or less.

It's easy to see this as a kind of failure. Who would want to be a publisher with only a dozen readers? It's also easy to see why the audience for most user-generated content is so small, filled as it is with narrow, spelling-challenged observations about going to the mall and picking out clothes for

Dixon. And it's easy to deride this sort of thing as self-absorbed publishing—why would anyone put such drivel out in public?

It's simple. They're not talking to you.

We misread these seemingly inane posts because we're so unused to seeing written material in public that isn't intended for us. The people posting messages to one another in small groups are doing a different kind of communicating than people posting messages for hundreds or thousands of people to read. More is different, but less is different too. An audience isn't just a big community; it can be more anonymous, with many fewer ties among users. A community isn't just a small audience either; it has a social density that audiences lack. The bloggers and social network users operating in small groups are part of a community, and they are enjoying something analogous to the privacy of the mall. On any given day you could go to the food court in a mall and find a group of teenagers hanging out and talking to one another. They are in public, and you could certainly sit at the next table over and listen in on them if you wanted to. And what would they be saying to one another? They'd be saying, "I can't believe I missed you last night!!! Trac said you were TRASHED off your ASS!" They'd be doing something similar to what they are doing on LiveJournal or Xanga, in other words, but if you were listening in on their conversation at the mall, as opposed to reading their post, it would be clear that you were the weird one.

Most user-generated content isn't "content" at all, in the sense of being created for general consumption, any more than a phone call between you and a relative is "family-generated content." Most of what gets created on any given day is just the ordinary stuff of life—gossip, little updates, thinking out loud—but now it's done in the same medium as professionally produced material. Similarly, people won't prefer professionally produced content in situations where community matters: I have a terrible singing voice, but my children would be offended if I played a well-sung version of "Happy Birthday" on the stereo, as opposed to singing it myself, badly.

Saying something to a few people we know used to be quite distinct from saying something to many people we don't know. The distinction between communications and broadcast media was always a function of technology rather than a deep truth about human nature. Prior to the internet, when we talked about media, we were talking about two different things: broadcast media and communications media. Broadcast media, such as radio and television but also newspapers and movies (the term refers to a message being broadly delivered from a central place, whatever the medium), are designed to put messages out for all to see (or in some cases, for all buyers or subscribers to see). Broadcast media are shaped, conceptually, like a megaphone, amplifying a one-way message from one sender to many receivers. Communications media, from telegrams to phone calls to faxes, are designed to facilitate two-way conversations. Conceptually, communications media are like a tube; the message put into one end is intended for a particular recipient at the other end.

Communications media was between one sender and one recipient. This is a one-to-one pattern—I talk and you listen, then you talk and I listen. Broadcast media was between one sender and many recipients, and the recipients couldn't talk back. This is a one-to-many pattern—I talk, and talk, and talk, and all you can do is choose to listen or tune out. The pattern we *didn't* have until recently was many-to-many, where communications tools enabled group conversation. E-mail was the first really simple and global tool for this pattern (though many others, like text messaging and IM, have since been invented).

Now that our communications technology is changing, the distinctions among those patterns of communication are evaporating; what was once a sharp break between two styles of communicating is becoming a smooth transition. Most user-generated content is created as communication in small groups, but since we're so unused to communications media and broadcast media being mixed together, we think that everyone is now broadcasting. This is a mistake. If we listened in on other phone calls, we'd know to expect small talk, inside jokes, and the like, but people's phone calls aren't out in the open. One of the driving forces behind much user-generated content is that conversation is no longer limited to social cul-de-sacs like the phone.

The distinction between broadcast and communications, which is to say between one-to-many and one-to-one tools, used to be so clear that we could distinguish between a personal and impersonal message just by the type of medium used. Someone writing you a letter might say "I love you," and someone on TV might say "I love you," but you would have no trouble understanding which of those messages was really addressed to you. We place considerable value on messages that are addressed to us personally, and we are good at distinguishing between messages meant for us individually (like love letters) and those meant for people like us (like those coming from late-night preachers and pitchmen). An entire industry, direct mail, sprang up around trying to trick people into believing that mass messages were really addressed to them personally. Millions of dollars have been spent on developing and testing ways of making bulk advertisements look like personal mail, including addressing the recipient by name and printing what looks like handwritten memos from the nominal sender. My annoyance at getting mail exhorting someone named Caly Shinky to "Act now!" comes from recognizing this trick while seeing it fail. Home shopping television shows use a related trick, instructing their phone sales representatives to be friendly to the callers and to compliment them on their good taste in selecting whatever it is they are buying, because they know that at least some of the motivation to buy comes from a desire to alleviate the loneliness of watching television. Though this friendliness makes each call take longer on average, it also makes the viewer happy, even though the original motivation to call came from watching people on TV—people who cannot, by definition, care about you personally.

Some user-generated content, of course, is quite consciously addressed to the public. Popular weblogs like Boing Boing (net culture), the Huffington Post (left-wing U.S. politics), and Power Line (right-wing U.S. politics) are all recognizably media outlets, with huge audiences instead of small clusters of friends. But between the small readership of the volleyball-playing Angel_An_Of_Lips on Xanga and the audience of over a million for Boing Boing, there is no obvious point where a blog (or indeed any user-created material) stops functioning like a diary for friends and starts functioning like a media outlet. Alisara Chirapongse (aka gnarlykitty) wrote about things of interest to her and her fellow Thai fashionistas, and then, during the coup, she briefly became a global voice. Community now shades into audience; it's as if your phone could turn into a radio station at the turn of a knob.

The real world affords us many ways of keeping public, private, and secret utterances separate from one another, starting with the fact that groups have until recently largely been limited to meeting in the real world, and things you say in the real world are heard only by the people you are talking to and only while you are talking to them. Online, by contrast, the default mode for many forms of communication is instant, global, and nearly permanent. In this world the private register suffers—those of us who grew up with a strong separation between communication and broadcast media have a hard time seeing something posted to a weblog as being in a private register, even when the content is obviously an in-joke or ordinary gossip, because we assume that if something is out where we can find it, it must have been written for us.

The fact that people are all talking to one another in these small clusters also explains why bloggers with a dozen readers don't have a small audience: they don't have an audience at all, they just have friends. In fact, as blogging was getting popular at the beginning of this decade, the blogging software with the most loyal users was none other than LiveJournal, which had more clusters of friends blogging for one another than any other blogging tool. If blogging were primarily about getting a big audience, LiveJournal should have suffered the most from disappointed users abandoning the service, but the opposite was the case. Writing things for your friends to read and reading what your friends write creates a different kind of pleasure than writing for an audience. Before the internet went mainstream, it took considerable effort to say something that would be heard by a significant number of people, so we regard any publicly available material as being offered directly to us. Now that the cost of posting things in a global medium has collapsed, much of what gets posted on any given day is in public but not for the public.

## Fame Happens

It's also possible to make the opposite mistake: not that conversational utterances are publishing, but that all publications are now part of a conversation. This view is common, though, and is based on the obvious notion that the Web is different from broadcast media like TV because the Web can support real interaction among users.

In this view, the effects of television are mainly caused by its technological limits. Television has millions of inbound arrows—viewers watching the screen—and no outbound arrows at all. You can see Oprah; Oprah can't see you. On the Web, by contrast, the arrows of attention are all potentially reciprocal; anyone can point to anyone else, regardless of geography, infrastructure, or other limits. If Oprah had a weblog, you could link to her, and she could link to you. This potential seems as if it should allow everyone to interact with everyone else, undoing the one-way nature of television. But calling that potential interactivity would be like calling a newspaper inter-active because it publishes letters to the editor.

The Web makes interactivity technologically possible, but what technology giveth, social factors taketh away. In the case of the famous, any potential interactivity is squashed, because fame isn't an attitude, and it isn't technological artifact. Fame is simply an imbalance between inbound and outbound atten-tion, more arrows pointing in than out. Two things have to happen for someone to be famous, neither of them related to technology. The first is scale: he or she has to have some min-imum amount of attention, an audience in the thousands or more. (This is why the internet version of the Warhol quote—"In the future everyone will be famous to fifteen people"—is appealing but wrong.) Second, he or she has to be unable to reciprocate. We know this pattern from television; audiences for the most popular shows are huge, and reciprocal attention is technologically impossible. We believed (often because we wanted to believe) that technical limits caused this imbalance in attention. When weblogs and other forms of interactive media began to spread, they enabled direct, unfiltered conver-sation among all parties and removed the structural imbal-ances of fame. This removal of the technological limits has exposed a second set of social ones.

Though the possibility of two-way links is profoundly good, it is not a cure-all. On the Web interactivity has no techno-logical limits, but it does still have strong cognitive limits: no matter who you are, you can only read so many weblogs, can trade e-mail with only so many people, and so on. Oprah has e-mail, but her address would become useless the minute it became public. These social constraints mean that even when a medium is two-way, its most popular practitioners will be forced into a one-way pattern. Whether Oprah *wants* to talk to each and every member of her audience is irrelevant: Oprah *can't* talk to even a fraction of a percent of her audience, ever, because she is famous, which means she is the recipient of more attention than she can return in any medium. These social constraints didn't much matter at small scale. In the early days of weblogs (prior to 2002, roughly) there was a remarkable and loose-jointed conversation among webloggers of all stripes, and those with a reasonable posting tempo could count themselves one of the party. In those days weblogging was mainly an interactive pursuit, and it happened so naturally that it was easy to imagine that interactivity was a basic part of the bargain.

Then things got urban, with millions of bloggers and readers. At this point social limits kicked in. If you have a weblog, and a thousand other webloggers point to you, you cannot read what they are saying, much less react. More is different: cities are not just large towns, and a big audience is not just a small one cloned many times. The limits on interaction that come with scale are hard to detect, because every visible aspect of the system stays the same. Nothing about the software or the users changes, but the increased population still alters the circumstances beyond your control. In this situation, no matter how assiduously someone wants to interact with their readers, the growing audience will ultimately defeat that possibility. Someone blogging alongside a handful of friends can read everything those friends write and can respond to any comments their friends make—the scale is small enough to allow for a real conversation. Someone writing for thousands of people, though, or millions, has to start choosing who to respond to and who to ignore, and over time, ignore becomes the default choice. They have, in a word, become famous.

Glenn Reynolds, a homegrown hero of the weblog world, reports over a million unique viewers a month for Instapundit .com, a circulation that would put him comfortably in the top twenty daily papers in the United States. You can see how interactivity is defeated by an audience of this size—spending even a minute a month interacting with just ten thousand of his readers (only one percent of his total audience) would take forty hours a week. This is what "interactivity" looks like at this scale—no interaction at all with almost all of the audience, and infrequent and minuscule interaction with the rest, and it has implications for media of all types. Weblogs won't destroy the one-way mirror of fame, and "interactive TV" is an oxymoron, because gathering an audience at TV scale defeats anything more interactive than voting for someone on *American Idol*.

The surprise held out by social tools like weblogs is that scale alone, even in a medium that allows for two-way connections, is enough to create and sustain the imbalance of fame. The mere technological possibility of reply isn't enough to overcome the human limits on attention. Charles Lindbergh couldn't bear to let anyone else answer his fan mail, promising himself he would get around to it eventually (which, of course, he never did). Egalitarianism is possible only in small social systems. Once a medium gets past a certain size, fame is a forced move. Early reports of the death of traditional media portrayed the Web as a kind of anti-TV—two-way where TV is one-way, interactive where TV is passive, and (implicitly) good where TV is bad. Now we know that the Web is not a perfect antidote to the problems of mass media, because some of those problems are human and are not amenable to technological fixes. This is bad news for that school of media criticism that has assumed that the authorities are keeping the masses down. In the weblog world there are no authorities, only masses, and yet the accumulated weight of attention continues to create the kind of imbalances we associate with traditional media.

The famous are different from you and me, because they cannot return or even acknowledge the attention they get, and technology cannot change that. If we want large systems where attention is unconstrained, fame will be an inevitable by-product, and as our systems get larger, its effects will become more pronounced, not less. A version of this is happening with e-mail—because it is easier to ask a question than to answer it, we get the curious effect of a group of people all able to overwhelm one another by asking, cumulatively, more questions than they can cumulatively answer. As Merlin Mann, a software usability expert, describes the pattern:

> Email is such a funny thing. People hand you these single little messages that are no heavier than a river pebble. But it doesn't take long until you have acquired a pile of pebbles that's taller than you and heavier than you could ever hope to move, even if you wanted to do it over a few dozen trips. But for the person who took the time to hand you their pebble,

> it seems outrageous that you can't handle that one
> tiny thing. "What 'pile'? It's just a pebble!"

E-mail, and particularly the ability to create group conversations
effortlessly without needing the permission of the recipients, is
providing a way for an increasing number of us to experience
the downside of fame, which is being unable to reciprocate in
the way our friends and colleagues would like us to.

The limiting effect of scale on interaction is bad news for
people hoping for the dawning of an egalitarian age ushered
in by our social tools. We can hope that fame will become
more dynamic, and that the elevation to fame will be more
bottom-up, but we can no longer hope for a world where everyone can interact with everyone else. Whatever the technology, our social constraints will mean that the famous of the
world will always be with us. The people with too much inbound attention live in a different environment from everyone
else; to paraphrase F. Scott Fitzgerald, the attention-rich are
different from you and me, in ways that are not caused by the
media they use, and in ways that won't go away even when
new media arrive.

For the last fifty years the two most important communications media in most people's lives were the telephone and
television: different media with different functions. It turns
out that the difference between conversational tools and
broadcast tools was arbitrary, but the difference between conversing and broadcasting is real. Even in a medium that allowed for perfect interactivity for all participants (something
we have a reasonable approximation of today), the limits of
human cognition will mean that scale alone will kill conversa-

tion. In such a medium, even without any professional bottle-necks or forced passivity, fame happens.

## Filtering as a Tool for Communities of Practice

Comparisons between the neatness of traditional media and the messiness of social media often overlook the fact that the comparison isn't just between systems of production but between systems of filtering as well. You can see how critical filtering tools are to the traditional landscape if you imagine taking a good-sized bookstore, picking it up, and shaking its contents out onto a football field. Somewhere in the resulting pile of books lie the works of Aristotle, Newton, and Auden, but if you wade in and start picking up books at random, you're much likelier to get *Love's Tender Fury* and *Chicken Soup for the Hoosier Soul*. We're so used to the way a bookstore is laid out that we don't notice how much prior knowledge we need to have about its layout and categories for it to be even minimally useful. As the investor Esther Dyson says, "When we call something intuitive, we often mean familiar."

The hidden contours of the filtering problem shaped much of what is familiar about older forms of media. Television shows, for instance, come in units of half an hour, not because the creators of television discovered that that is the aesthetically ideal unit of time, but because audiences had to remember when their favorite shows were on. A show that starts at

7:51 and goes on until 8:47 is at a considerable disadvantage to a show that starts at 8:00 and goes till 9:00, and that disadvantage is entirely cognitive—the odd times are simply harder to remember. (It's hard to have appointment TV if you can't recall when the appointment is.) The length and time slots of television had nothing to do with video as a medium and everything to do with the need to aid the viewer's memory. Similarly, everything from *TV Guide* to the rise of content-specific channels on cable like MTV and the Cartoon Network were responses to the problem of helping viewers find their way to interesting material.

Traditional media have a few built-in constraints that make the filtering problem relatively simple. Most important, publishing and broadcasting cost money. Any cost creates some sort of barrier, and the high cost of most traditional media creates high barriers. As a result, there is an upper limit to the number of books, or television shows, or movies that can exist. Since the basic economics of publishing puts a cap on the overall volume of content, it forces every publisher or producer to filter the material in advance.Simply to remain viable, anyone producing traditional media has to decide what to produce and what not to; the good work has to be sorted from the mediocre in advance of publication.

Though the filtering of the good from the mediocre starts as an economic imperative, the public enjoys the value of that filtering as well, because we have historically relied on the publisher's judgment to help ensure minimum standards of quality. Where publishing is hard and expensive, every instance of the written word comes with an implicit promise: someone besides the writer thought this was worth reading. Every book and magazine article and newspaper (as well as every published photo and every bit of broadcast speech or song or bit of video) had to pass through some editorial judgment. You can see this kind of filtering at work whenever someone is referred to as a "published author." The label is a way of assuring people that some external filter has been applied to the work. (The converse of this effect explains our skepticism about self-published books and the label reserved for publishers who print such books—the vanity press.)

The old ways of filtering were neither universal nor ideal; they were simply good for the technology of the day, and reasonably effective. We were used to them, and now we have to get used to other ways of solving the same problem. Mass amateurization has created a filtering problem vastly larger than we had with traditional media, so much larger, in fact, that many of the old solutions are simply broken. The brute economic logic of allowing anyone to create anything and make it available to anyone creates such a staggering volume of new material, every day, that no group of professionals will be adequate to filter the material. Mass amateurization of publishing makes mass amateurization of filtering a forced move. Filter-then-publish, whatever its advantages, rested on a scarcity of media that is a thing of the past. The expansion of social media means that the only working system is publish-then-filter.

We have lost the clean distinctions between communications media and broadcast media. As social media like MySpace now scale effortlessly between a community of a few and an audience of a few million, the old habit of treating communications tools like the phone differently from broadcast tools like television no longer makes sense. The two patterns shade into each other, and now small group communications and large broadcast outlets all exist as part of a single interconnected ecosystem. This change is the principal source of "user-generated content." Users—people—have always talked to one another, incessantly and at great length. It's just that the user-to-user messages were kept separate from older media, like TV and newspapers.

The activities of the amateur creators are self-reinforcing. If people can share their work in an environment where they can also converse with one another, they will begin talking about the things they have shared. As the author and activist Cory Doctorow puts it, "Conversation is king. Content is just something to talk about." The conversation that forms around shared photos, videos, weblog posts, and the like is often about how to do it better next time—how to be a better photographer or a better writer or a better programmer. The goal of getting better at something is different from the goal of being good at it; there is a pleasure in improving your abilities even if that doesn't translate into absolute perfection. (As William S. Burroughs, the Beat author, once put it, "If a thing is worth doing, it's worth doing badly.") On Flickr, many users create "high dynamic range" photos (HDR), where three exposures of the same shot are combined. The resulting photos are often quite striking, as they have a bigger range of contrast—the brights are brighter and the darks are darker—than any of the individual source photos. Prior to photo-sharing services, anyone looking at such a photo could wonder aloud, "How did they do that?" With photo sharing, every picture is a potential site for social interaction, and viewers can and do ask the question directly, "How did you do that?," with a real hope of getting an answer. The conversations attached to these photos are often long and detailed, offering tutorials and advice on the best tools and techniques for creating HDR photos. This form of communication is what the sociologist Etienne Wenger calls a community of practice, a group of people who converse about some shared task in order to get better at it.

John Seely Brown and Paul Duguid, in their book *The Social Life of Information*, put the dilemma this way: "What if HP [Hewlett-Packard] knew what HP knows?" They had observed that the sum of the individual minds at HP had much more information than the company had access to, even though it was allowed to direct the efforts of those employees. Brown and Duguid documented ways in which employees do better at sharing information with one another directly than when they go through official channels. They noticed that supposedly autonomous Xerox repair people were gathering at a local breakfast spot and trading tips about certain kinds of repairs, thus educating one another in the lore not covered by the manuals. Without any official support, the repair people had formed a community of practice. Seeing this phenomenon, Brown convinced Xerox to give the repair staff walkie-talkies, so they could continue that sort of communication during the day.

By lowering transaction costs, social tools provide a platform for communities of practice. The walkie-talkies make asking and answering "How did you do that?" questions easy. They would seem to transfer the burden from the asker to the answerer, but they also raise the answerer's status in the community. By providing an opportunity for the visible display of expertise or talent, the public asking of questions creates a motivation to answer in public as well, and that answer, once perfected, persists even if both the original asker and the answerer lose interest. Communities of practice are inherently cooperative, and are beautifully supported by social tools, because that is exactly the kind of community whose members can recruit one another or allow themselves to be found by interested searchers. They can thrive and even grow to enormous size without advertising their existence in public. On Flickr alone there are thousands of groups dedicated to exploring and perfecting certain kinds of photos: landscape and portraiture, of course, but also photos featuring the color red, or those composed of a square photo perfectly framing a circle, or photos of tiny animals clinging to human fingers.

There are thousands of examples of communities of practice. The Web company Yahoo hosts thousands of mailing lists, many of them devoted to advancing the practice of everything from Creole cooking to designing radio-controlled sailboats. Gaia Online is a community for teenage fans of anime and manga, the Japanese animation and cartoon forms; their discussion groups include long threads devoted to critiquing one another's work and tutorials on the arcana of the form, like how to draw girls with really big eyes. Albino Blacksheep is a community for programmers working on interactive games and animation. All these groups offer the

kind of advice, feedback, and encouragement that character-
izes communities of practice. These communities can be
huge—Gaia Online has millions of users. For most of the
history of the internet, online groups were smaller than traditional audiences—big-city newspapers and national TV
shows reached more people than communal offerings. Now,
though, with a billion people online and more on the way, it's
easy and cheap to get the attention of a million people or,
more important, to help those people get one another's attention. In traditional media we know the names of most of the
newspapers that have more than a million readers, because
they have to appeal to such a general audience, but sites like
Albino Blacksheep and Gaia Online occupy the odd and new
category of meganiches—nichelike in their appeal to a very
particular audience, but with a number of participants previously available only to mainstream media.

Every webpage is a latent community. Each page collects
the attention of people interested in its contents, and those
people might well be interested in conversing with one another, too. In almost all cases the community will remain
latent, either because the potential ties are too weak (any
two users of Google are not likely to have much else in common) or because the people looking at the page are separated by too wide a gulf of time, and so on. But things like
the comments section on Flickr allow those people who do
want to activate otherwise-latent groups to at least try it. The
basic question "How did you do that?" seems like a simple
request for a transfer of information, but when it takes place
out in public, it is also a spur to such communities of prac-

tice, bridging the former gap between publishing and conversation.

Though some people participate in communities of practice for the positive effects on their employability, within the community they operate with different, nonfinancial motives. Love has profound effects on small groups of people—it helps explain why we treat our family and friends as we do—but its scope is local and limited. We feed our friends, care for our children, and delight in the company of loved ones, all for reasons and in ways that are impossible to explain using the language of getting and spending. But large-scale and long-term effort require that someone draw a salary. Even philanthropy exhibits this property; the givers can be motivated by a desire to do the right thing, but the recipient, whether the Red Cross or the Metropolitan Opera, has to have a large staff to direct those donations toward the desired effect. Life teaches us that motivations other than getting paid aren't enough to add up to serious work.

And now we have to unlearn that lesson, because it is less true with each passing year. People now have access to myriad tools that let them share writing, images, video—any form of expressive content, in fact—and use that sharing as an anchor for community and cooperation. The twentieth century, with the spread of radio and television, was the broadcast century. The normal pattern for media was that they were created by a small group of professionals and then delivered to a large group of consumers. But media, in the word's literal sense as the middle layer between people, have always been a three-part affair. People like to consume media, of course, but they also like to produce it ("Look what I made!") and they like to share it ("Look what I found!"). Because we now have media that support both making and sharing, as well as consuming, those capabilities are reappearing, after a century mainly given over to consumption. We are used to a world where little things happen for love and big things happen for money. Love motivates people to bake a cake and money motivates people to make an encyclopedia. Now, though, we can do big things for love.

## Revolution and Coevolution

There's a story in my family about my parents' first date. My father, wanting to impress my mother, decided to take her to a drive-in movie. Lacking anything to drive in to the drive-in, however, he had to borrow *his* father's car. Once they were at the movie, my mother, wanting to impress my father, ordered the most sophisticated drink available, which was a root beer float. Now my mother hates root beer, always has, and after imbibing it, she proceeded to throw up on the floor of my grandfather's car. My father had to drive her home, missing the movie he'd driven fifteen miles and paid a dollar to see. Then he had to clean the car and return it with an explanation and an apology. (There was, fortunately for me, a second date.)

Now, what part of that story is about the internal combustion engine? None of it, in any obvious way, but all of it, in another way. No engine, no cars. No cars, no using cars for dates. (The effect of automobiles on romance would be hard to overstate.) No dates in cars, no drive-in movies. And so on. Our life is so permeated with the automotive that we understand immediately how my father must have felt when my grandfather let him borrow the car, and how carefully he must have cleaned it before returning it, without thinking about internal combustion at all.

This pattern of coevolution of technology and society is true of communications tools as well. Here's a tech history question: which went mainstream first, the fax or the Web? People over thirty-five have a hard time understanding why you'd even ask—the fax machine obviously predates the Web for general adoption. Here's another: which went mainstream first, the radio or the telephone? The same people often have to think about this question, even though the practical demonstration of radio came almost two decades after that of the telephone, a larger gap than separated the fax and the Web. We have to think about radio and the telephone because for everyone alive today, those two technologies have always existed. And for college students today, that is true of the fax and the Web. Communications tools don't get socially interesting until they get technologically boring. The invention of a tool doesn't create change; it has to have been around long enough that most of society is using it. It's when a technology becomes normal, then ubiquitous, and finally so pervasive as to be invisible, that the really profound changes happen, and for young people today, our new social tools have passed normal and are heading to ubiquitous, and invisible is coming.

We are living in the middle of the largest increase in

expressive capability in the history of the human race. More people can communicate more things to more people than has ever been possible in the past, and the size and speed of this increase, from under one million participants to over one billion in a generation, makes the change unprecedented, even considered against the background of previous revolutions in communications tools. The truly dramatic changes in such tools can be counted on the fingers of one hand: the printing press and movable type (considered as one long period of innovation); the telegraph and telephone; recorded content (music, then movies); and finally the harnessing of radio signals (for broadcasting radio and TV). None of these examples was a simple improvement, which is to say a better way of doing what a society already did. Instead, each was a real break with the continuity of the past, because any radical change in our ability to communicate with one another changes society.

There was a persistent imbalance in these earlier changes, however. The telephone, the technological revolution that put the most expressive power in the hands of the individual, didn't create an audience; telephones were designed for conversation. Meanwhile the printing press and recorded and broadcast media created huge audiences but left control of the media in the hands of a small group of professionals. As mobile phones and the internet both spread and merge, we now have a platform that creates both expressive power and audience size. Every new user is a potential creator and consumer, and an audience whose members can cooperate directly with one another, many to many, is a former audience. Even if what the audience creates is nothing more than a few text messages or e-mails, those messages can be addressed not just to individuals but to groups, and they can be copied and forwarded endlessly.

Our social tools are not an improvement to modern society; they are a challenge to it. A culture with printing presses is a different *kind* of culture from one that doesn't have them. New technology makes new things possible: put another way, when new technology appears, previously impossible things start occurring. If enough of those impossible things are important and happen in a bundle, quickly, the change becomes a revolution.

The hallmark of revolution is that the goals of the revolutionaries cannot be contained by the institutional structure of the existing society. As a result, either the revolutionaries are put down, or some of those institutions are altered, replaced, or destroyed. We are plainly witnessing a restructuring of the media businesses, but their suffering isn't unique, it's prophetic. All businesses are media businesses, because whatever else they do, all businesses rely on the managing of information for two audiences—employees and the world. The increase in the power of both individuals and groups, outside traditional organizational structures, is unprecedented. Many institutions we rely on today will not survive this change without significant alteration, and the more an institution or industry relies on information as its core product, the greater and more complete the change will be.

The linking of symmetrical participation and amateur production makes this period of change remarkable. Symmetrical participation means that once people have the capacity to receive information, they have the capability to send it as well.

Owning a television does not give you the ability to make TV shows, but owning a computer means that you can create as well as receive many kinds of content, from the written word through sound and images. Amateur production, the result of all this new capability, means that the category of "consumer" is now a temporary behavior rather than a permanent identity.

# CHAPTER 5

# PERSONAL MOTIVATION MEETS COLLABORATIVE PRODUCTION

*Collaborative production, where people have to coordinate with one another to get anything done, is considerably harder than simple sharing, but the results can be more profound. New tools allow large groups to collaborate, by taking advantage of nonfinancial motivations and by allowing for wildly differing levels of contribution.*

Perhaps the most famous example of distributed collaboration today is Wikipedia, the collaboratively created encyclopedia that has become one of the most visited websites in the world. Jimmy Wales and Larry Sanger founded Wikipedia in 2001 as an experimental offshoot of their original idea, a free online encyclopedia of high quality called Nupedia. Nupedia was to be written, reviewed, and managed by experts volunteering their time. Wales had had a taste of collaboratively produced work while running Bomis, an internet company he'd helped found in 1996. Bomis was in the business of helping (mainly male) users create and show collections of related websites on subjects like overengineered cars and underdressed starlets; it was like a user-curated Maxim. He had seen how quickly and cheaply the users could share information with one another, and he thought that sort of collaborative creation could be applied to other domains. He sketched out the idea for Nupedia, secured investment from Bomis in early 2000, and hired Sanger, a Ph.D. candidate in philosophy who shared Wales's interest in theories of knowledge, as employee number one.

Sanger began designing a process for creating Nupedia articles, and after several weeks of preparation, he and Wales announced the project with a stirring question:

> Suppose scholars the world over were to learn of a serious online encyclopedia effort in which the results were not proprietary to the encyclopedists, but were freely distributable . . . in virtually any desired medium. How quickly would the encyclopedia grow?

Not very quickly, as it turned out. Nine months after that announcement, Wales and Sanger's big idea wasn't working; if scholars the world over had learned of Nupedia, they certainly hadn't responded by rushing in to help. In the months after the original announcement, most of the effort had been spent on recruiting a volunteer advisory board and on establishing editorial policy guidelines and a process for the creation, review, revision, and publication of articles. This process, intended to set a minimum standard of quality, had also set a maximum rate of progress: slow. At the end of that gestation period there were fewer than twenty finished articles and another few in various stages of work. (One can't call them stages of completion, since completion was something Nupedia was visibly bad at.)

For those scholars who were successfully recruited to participate, the flow of work from a draft article to something published involved seven separate steps. If an article was stopped at any of those steps—for review, fact-checking, spell-checking, whatever—it could remain stopped indefinitely. Increasingly frustrated with the slow pace, and aware that their own process had erected many new barriers to replace the ones the Web had removed, Sanger suggested a new strategy to Wales: use a tool called a wiki to create the first draft of Nupedia articles.

The first wiki was created by Ward Cunningham, a software engineer, in 1995. (The name wiki is taken from the Hawaiian word for "quick.") Cunningham wanted a way for the software community to create a repository of shared design wisdom. He observed that most of the available tools for collaboration were concerned with complex collections of roles and requirements—only designated writers could create text, whereas only editors could publish it, but not until proofreaders had approved it, and so on. Cunningham made a different, and radical, assumption: groups of people who want to collaborate also tend to trust one another. If this was true, then a small group could work on a shared writing effort without needing formal management or process.

Cunningham's wiki, the model for all subsequent wikis, is a user-editable website. Every page on a wiki has a button somewhere, usually reading "Edit this," that lets the reader add, alter, or delete the contents of the page. With a book or a magazine the distinction between reader and writer is enforced by the medium; with a wiki someone can cross back and forth between the two roles at will. (Flexibility of role is a common result of mass amateurization.) Whenever a user edits any-thing on a given webpage, the wiki records the change and saves the previous version. Every wiki page is thus the sum total of accumulated changes, with all earlier edits stored as historical documentation. This was a gamble, but Cunningham's design worked beautifully; the first wiki, called the Portland Pattern Repository, became an invaluable collection of software engineering wisdom without requiring either formal oversight or editorial controls. By placing the process in the hands of the users rather than embedding it in the tool, the wiki dispensed with the slowness that often comes with highly structured work environments. Seeing this effect, other groups began to adopt wikis.

In early 2001 a friend of Sanger's told him about wikis, and he in turn introduced the idea to Wales. They set up a test wiki on Nupedia as a way to create rough drafts, which had two immediate effects. First, it became much easier to create initial versions of articles. The second effect, which they had not an-ticipated, was swift and vehement objection from their own advisory board. The board had been recruited to oversee a rigor-ous process, designed and run by experts, and the wiki offended their sense of the mission. A few days after they'd launched it, Wales and Sanger had to move the nascent wiki off the Nupedia site to placate the board. As the wiki now needed its own URL, they chose Wikipedia.com, and Wikipedia was born.

Once Wikipedia was up, Sanger posted a note to the Nupedia mailing list, which by that point had about two thousand members, saying, "Humor me. Go there and add a little article. It will take all of five or ten minutes." The change was immediate and dramatic; Wikipedia surpassed Nupedia in total number of articles in its first few weeks of existence. By the end of the year, with fifteen thousand articles in place and the rate of growth continuing to increase, two things became clear: Wikipedia was viable, and Nupedia was not.

Seeing this success, Sanger shifted to the Wikipedia effort, dropping his Nupedia "editor in chief" title along the way and instead calling himself "Chief Organizer." Despite the mollifying nature of this title, he managed to infuriate the other participants when he said, in a message to the Wikipedia mailing list, "I do reserve the right to permanently delete things—particularly when they have little merit and when they are posted by people whose main motive is evidently to undermine my authority and therefore, as far as I'm concerned, damage the project." Sanger's assumption of special rights over the project, and his equation of those prerogatives with the project's success, only worsened the friction about his role.

In part because of these clashes, and in part because Wikipedia's growth neither created nor required revenue, Sanger was laid off at the end of 2001. The Wikipedia project was later transferred to Wikipedia.org to cement its nonprofit status; the progression from Nupedia to Wikipedia as we know it today was complete. Continued growth was uninterrupted by Sanger's departure; Wikipedia has continued to grow steadily in both articles and users. The English-language version crossed the two-million-article mark in September

2007. The English-language Wikipedia is the only noncommercial site in the top twenty Web sites for the United States.

## Wikipedia's Content

Mere volume would be useless if Wikipedia articles weren't any good, however. By way of example, the article on Pluto as of May 2007 begins:

> Pluto, also designated 134340 Pluto, is the second-largest known dwarf planet in the Solar System and the tenth-largest body observed directly orbiting the Sun. Originally considered a planet, Pluto has since been recognized as the largest member of a distinct region called the Kuiper belt. Like other members of the belt, it is primarily composed of rock and ice and is relatively small; approximately a fifth the mass of the Earth's Moon and a third its volume. It has an eccentric orbit that takes it from 29 to 49 AU from the Sun, and is highly inclined with respect to the planets. As a result, Pluto occasionally comes closer to the Sun than the planet Neptune.

That paragraph includes ten links to other Wikipedia articles on the solar system, astronomical units (AU), and so on. The article goes on for five thousand words and ends with an extensive list of links to other sites with information about Pluto. This kind of thing—a quick overview, followed by broad and sometimes quite lengthy descriptions, ending with pointers to more information—is pretty much what you'd want in an encyclopedia.

The Pluto article is not unusual; you can find articles of similarly high quality all over the site.

> The Okeechobee Hurricane, or Hurricane San Felipe Segundo, was a deadly hurricane that struck the Leeward Islands, Puerto Rico, the Bahamas, and Florida in September of the 1928 Atlantic hurricane season. It was the first recorded hurricane to reach Category 5 status on the Saffir-Simpson Hurricane Scale in the Atlantic basin.

or

> Ludwig Josef Johann Wittgenstein (April 26, 1889 in Vienna, Austria—April 29, 1951 in Cambridge, England) was an Austrian philosopher who contributed several ground-breaking ideas to philosophy, primarily in the foundations of logic, the philosophy of mathematics, the philosophy of language, and the philosophy of mind. His influence has been wide-ranging, placing him among the most significant philosophers of the 20th century.

And so on. There are hundreds of thousands of articles whose value is both relied on and improved daily.

The most common criticism of Wikipedia over the years stemmed from simple disbelief: "That can't work." Sanger understood this objection and titled an early essay on the growth of Wikipedia "Wikipedia is wide open. Why is it growing so fast? Why isn't it full of nonsense?" In that article he ascribed at least part of the answer to group editing:

> Wikipedia's self-correction process (Wikipedia co-founder Jimmy Wales calls it "self-healing") is very robust. There is considerable value created by the public review process that is continually ongoing on Wikipedia—value that is very easy to underestimate, for those who have not experienced it adequately.

One other fateful choice, which actually predates the founding of Wikipedia itself, was the name, or rather the "-pedia" suffix. Wikipedia, like all social tools, is the way it is in part because of the way the software works and in part because of the way the community works. Though wikis can be used for many kinds of writing, the early users were guided by the rhetorical models of existing encyclopedias, which helped synchronize the early work: there was a shared awareness of the kind of writing that should go into a project called Wikipedia. This helped coordinate the users in ways that were not part of the software but were part of the community that used the software.

Wikipedia has now transcended the traditional functions of an encyclopedia. Within minutes of the bombs going off in the London transit system, someone created a Wikipedia page called "7 July 2005 London bombings." The article's first incarnation was five sentences long and attributed the explosions to a power surge in the Underground, one of the early theories floated before the bus bombing was linked to the Underground explosions. The Wikipedia page received more than a thousand edits in its first four hours of existence, as additional news came in; users added numerous pointers to traditional news sources (more symbiosis) and a list of contact numbers for people either trying to track loved ones or simply figuring out how to get home. What was conceived as an open encyclopedia in 2001 has become a general-purpose tool for gathering and distributing information quickly, a use that further cemented Wikipedia in people's mind as a useful reference work. Note the virtuous circle at work here: because enough people thought of using Wikipedia as a coordinating resource, it became one, and because it became one, more people learned to think of it as a coordinating resource. This evolution was made possible precisely because the community had gotten the narrower version of an encyclopedia right earlier, which provided a high-visibility platform for further experimentation.

Skepticism about Wikipedia's basic viability made some sense back in 2001; there was no way to predict, even with the first rush of articles, that the rate of creation and the average quality would both remain high, but today those objections have taken on the flavor of the apocryphal farmer beholding his first giraffe and exclaiming, "Ain't no such animal!" Wikipedia's daily utility for millions of users has been settled; the interesting questions are elsewhere.

## Unmanaged Division of Labor

It's easy to understand how Cunningham's original wiki functioned; a small group that knows one another presents organizational challenges no worse than getting a neighborhood poker game going. But Wikipedia doesn't operate at the scale of a neighborhood poker game; it operates at the scale of a Vegas casino. Something this big seems like it should require managers, a budget, a formal work-flow process. Without those things how could it possibly work? The simple but surprising answer is: spontaneous division of labor. Division of labor is usually associated with highly managed settings, but it's implemented here in a far more unmanaged way. Wikipedia is able to aggregate individual and often tiny contributions, hundreds of millions of them annually, made by millions of contributors, all performing different functions.

Here's how it works. Someone decides that an article on, say, asphalt should exist and creates it. The article's creator doesn't need to know everything (or indeed much of anything) about asphalt. As a result, such articles often have a "well, duh" quality to them. The original asphalt article read, in full, "Asphalt is a material used for road coverings." The article was created in March 2001, at the dawn of Wikipedia, by a user named Cdani, as little more than a placeholder saying, "We should have an article on asphalt here." (Wikipedians call this a "stub.")

Once an article exists, it starts to get readers. Soon a self-selecting group of those readers decide to become contributors. Some of them add new text, some edit the existing article, some add references to other articles or external sources, and some fix typos and grammatical errors. None of these people needs to know everything about asphalt; all contributions can be incremental. And not all edits are improvements: added material can clutter a sentence, intended corrections can unintentionally introduce new errors, and so on. But every edit is itself provisional. This works to Wikipedia's benefit partly because bad changes can be rooted out faster, but also partly because human knowledge is provisional. During 2006 a debate broke out among astronomers on whether to consider Pluto a planet or to relegate it to another category; as the debate went on, Wikipedia's Pluto page was updated to reflect the controversy, and once Pluto was demoted to the status of "dwarf planet," the Pluto entry was updated to reflect that almost immediately.

A Wikipedia article is a process, not a product, and as a result, it is never finished. For a Wikipedia article to improve, the good edits simply have to outweigh the bad ones. Rather than filtering contributions before they appear in public (the process that helped kill Nupedia), Wikipedia assumes that new errors will be introduced less frequently than existing ones will be corrected. This assumption has proven correct; despite occasional vandalism, Wikipedia articles get better, on average, over time.

It's easy to understand division of labor in industrial settings. A car comes into being as it passes down an assembly from one group of specialists to the next—first the axle, then the wheels. A wiki's division of labor is nothing like that. By 2007, the asphalt article has had 129 different contributors, who have subdivided it into two separate articles, one on asphalt, the petroleum derivative, and another on asphalt concrete, the road covering. To each of these articles, the contributors have added or edited sections on the chemistry, history, and geographic distribution of asphalt deposits, on different types of asphalt road surfaces, and even on the etymology of the word "asphalt," transforming the original seven-word entry into a pair of detailed and informative articles. No one person was responsible for doing or even managing the work, and yet researching, writing, editing, and proofreading have all unfolded over the course of five years. This pattern also exists across Wikipedia as a whole: one person can write new text on asphalt, fix misspellings in Pluto, and add external references for Wittgenstein in a single day. This system also allows great variability of effort—of the 129 contributors on the subject of asphalt, a hundred of them contributed only one edit each, while the half-dozen most active editors contributed nearly fifty edits among them, almost a quarter of the total. The most active contributor on the subject of asphalt, a user going by SCEhardt, is ten times more active than the average contributor and over a hundred times more active than the least active contributor.

This situation is almost comically chaotic—a car company would go out of business in weeks if it let its workers simply work on what they wanted to, when they wanted to. A car company has two jobs. The obvious one is making cars, but the other job is being a company. It's hard work to be a company; it requires a great deal of effort and a great deal of predictability. The inability to count on an employee's particular area of expertise, or even on their steady presence, would doom such an enterprise from the start. There is simply no commercially viable way to let employees work on what interests them as the mood strikes. There is, however, a noncommercial way to do so, which involves being effective without worrying about being efficient.

Wikis avoid the institutional dilemma. Because contributors aren't employees, a wiki can take a staggering amount of input with a minimum of overhead. This is key to its success: it does not need to make sure its contributors are competent, or producing steadily, or even showing up. Mandated specialization of talent and consistency of effort, seemingly the hallmarks of large-scale work, actually have little to do with division of labor itself. A business needs employee A and employee B to put in the same effort if they are doing the same job, because it needs interchangeability and because it needs to reduce friction between energetic and lazy workers. By this measure, most contributors to Wikipedia are lazy. The majority of contributors edit only one article, once, while the majority of the effort comes from a much smaller and more active group. (The two asphalt articles, with a quarter of the work coming from six contributors, are a microcosm of this general phenomenon.) Since no one is being paid, the energetic and occasional contributors happily coexist in the same ecosystem.

The freedom of contributors to jump from article to article and from task to task makes the work on any given article unpredictable, but since there are no shareholders or managers or even customers, predictability of that sort doesn't matter. Furthermore, since anyone can act, the ability of the people in charge to kill initiatives through inaction is destroyed. This is what befell Nupedia; because everyone working on that project understood that only experts were to write articles, no one would even begin an article they knew little about, and as long as the experts did nothing (which, on Nupedia, is mostly what they did), nothing happened. In an expert-driven system, an article on asphalt that read "Asphalt is a material used for road coverings" would never appear, even as a stub. So short! So uninformative! Why, anyone could have written that! Which, of course, is the principal advantage of Wikipedia.

In a system where anyone is free to get something started, however badly, a short, uninformative article can be the anchor for the good article that will eventually appear. Its very inadequacy motivates people to improve it; many more people are willing to make a bad article better than are willing to start a good article from scratch. In 1991 Richard Gabriel, a software engineer at Sun Microsystems, wrote an essay that included a section called "Worse Is Better," describing this effect. He contrasted two programming languages, one elegant but complex versus another that was awkward but simple. The belief at the time was that the elegant solution would eventually triumph; Gabriel instead predicted, correctly, that the language that was simpler would spread faster, and as a result, more people would come to care about improving the simple language than improving the complex one. The early successes of a simple model created exactly the incentives (attention, the desire to see your work spread) needed to create serious improvements. These kinds of incentives help ensure that, despite the day-to-day chaos, a predictable pattern emerges over time: readers continue to read, some of them become contributors, Wikipedia continues to grow, and articles continue to improve. The process is more like creating a coral reef, the sum of millions of individual actions, than creating a car. And the key to creating those individual actions is to hand as much freedom as possible to the average user.

## A Predictable Imbalance

Anything that increases our ability to share, coordinate, or act increases our freedom to pursue our goals in congress with one another. Never have so many people been so free to say and do so many things with so many other people. The free-dom driving mass amateurization removes the technological obstacles to participation. Given that everyone now has the tools to contribute equally, you might expect a huge increase in equality of participation. You'd be wrong.

You may have noticed a great imbalance of participation in many examples in this book. The Wikipedia articles for as-phalt had 129 contributors making 205 total edits, but the bulk of the work was contributed by a small fraction of par-ticipants, and just six accounted for about a quarter of the edits. A similar pattern appears on Flickr: 118 photographers contributed over three thousand Mermaid Parade photos to Flickr, but the top tenth contributed half of those, and the most active photographer, Czarina, contributed 238 photos (about one in twelve) working alone. This shape, called a power law distribution, is shown in Figure 5-1.

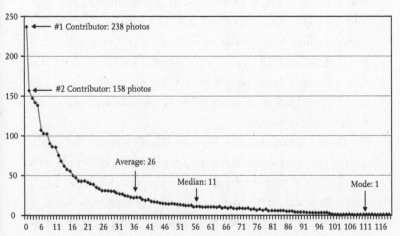

**Figure 5-1:** The distribution of photographers contributing photos of the 2005 Coney Island Mermaid Parade.

Five points are shown on this graph. The two leftmost data points are the most and second-most active photographers. The most active photographer is far more active than the second most active, and they are both far more active than most of the rest of the photographers. The average number of photos taken (all photos divided among all photographers) is twenty-six, while the median (the middle photographer) took eleven photos, and the mode (the number of photos that appeared most frequently) is a single photo.

Note the sharp drop-off in the number of photos between the top few contributors and most of the participants. Notice too that because of the disproportionate contributions of these few photographers, three-quarters of the photographers contributed a below-average number of pictures. This pattern is general to social media: on mailing lists with more than a couple dozen participants, the most active writer is generally much more active than the person in the number-two slot, and far more active than average. The longest conversation goes on much longer than the second-longest one, and much longer than average, and so on. Bloggers, Wikipedia contributors, photographers, people conversing on mailing lists, and social participation in many other large-scale systems all exhibit a similar pattern.

There are two big surprises here. The first is that the imbalance is the same shape across a huge number of different kinds of behaviors. A graph of the distribution of photo labels (or "tags") on Flickr is the same shape as the graph of readers-per-weblog and contributions-per-user to Wikipedia. The general form of a power law distribution appears in social settings when some set of items—users, pictures, tags—is ranked by frequency of occurrence. You can rank a group of Flickr users by the number of pictures they submit. You can rank a collection of pictures by the number of viewers. You can rank tags by the number of pictures they are applied to. All of these graphs will be in the rough shape of a power law distribution.

The second surprise is that the imbalance drives large social systems rather than damaging them. Fewer than two percent of Wikipedia users ever contribute, yet that is enough to create profound value for millions of users. And among those contributors, no effort is made to even out their contributions. The spontaneous division of labor driving Wikipedia wouldn't be possible if there were concern for reducing inequality. On the contrary, most large social experiments are engines for harnessing inequality rather than limiting it. Though the word "ecosystem" is overused as a way to make simple situations seem more complex, it is merited here, because large social systems cannot be understood as a simple aggregation of the behavior of some nonexistent "average" user.

The most salient characteristic of a power law is that the imbalance becomes more extreme the higher the ranking. The operative math is simple—a power law describes data in which the nth position has 1/nth of the first position's rank. In a pure power law distribution, the gap between the first and second position is larger than the gap between second and third, and so on. In Wikipedia article edits, for example, you would expect the second most active user to have committed only half as many edits as the most active user, and the tenth most active to have committed one-tenth as many. This is the shape behind the so-called 80/20 rule, where, for example, 20 percent of a store's inventory accounts for 80 percent of its revenues, and it has been part of social science literature since Vilfredo Pareto, an Italian economist working in the early 1900s, found a power law distribution of wealth in every country he studied; the pattern was so common that he called it "a predictable imbalance." This is also the shape behind Chris Anderson's discussion in *The Long Tail;* most items offered at online retailers like iTunes and Amazon don't sell well, but in aggregate they generate considerable income. The pattern doesn't apply just to goods, though, but to social interactions as well. Real-world distributions are only an approximation of this formula, but the imbalance it creates appears in an astonishing number of places in large social systems.

No matter how you display it, this shape is very different from the bell curve distribution we are used to. Imagine going out into your community and measuring the height of two hundred men selected at random. For anything like height that falls on a bell curve, knowing any one of the numbers— average, median, or mode—is a clue to the others. If you know the height of the median man, or the most common height among all the men, you can make an educated guess about the average height. And most critically, whatever you know about the average height can be assumed to be most representative of the group.

Now imagine height were described not by a bell curve but by a power law. If the average height of two hundred men was five foot ten; the most frequent (or modal) height would be held by dozens of men who were each only a foot tall, the median height would be two feet tall (a hundred men shorter than two feet, and a hundred taller). Most important, in such a distribution, the five tallest men would be 40, 50, 66, 100, and 200 feet tall respectively. Height doesn't follow a power law (fortunately for the world's tailors and architects), but the distributions of many social systems do. The most active contributor to a Wikipedia article, the most avid tagger of Flickr photos, and the most vocal participant in a mailing list all tend to be much more active than the median participant, so active in fact that any measure of "average" participation becomes meaningless. There is a steep decline from a few wildly active participants to a large group of barely active participants, and though the average is easy to calculate, it doesn't tell you much about any given participant.

Any system described by a power law, where mean, median, and mode are so different, has several curious effects. The first is that, by definition, most participants are below average. This sounds strange to many ears, as we are used to a world where average means middle, which is to say where average is the same as the median. You can see this "below average" phenomenon at work in the economist's joke: Bill Gates walks into a bar, and suddenly everyone inside becomes a millionaire, on average. The corollary is that everyone else in the bar also acquires a below-average income. The other surprise of such systems is that as they get larger, the imbalance between the few and the many gets larger, not smaller. As we get more weblogs, or more MySpace pages, or more YouTube videos, the gap between the material that gets the most attention and merely average attention will grow, as will the gap between average and median.

You cannot understand Wikipedia (or indeed any large social system) by looking at any one user or even a small group and assuming they are representative of the whole. The most active few users account for a majority of the edits, even though they make up a minority, and often a tiny minority, of contributors. But even this small group does not account for the whole success of Wikipedia, because many of these active users are doing things like correcting typos or making small changes, while users making only one edit are sometimes adding much larger chunks of relevant information.

Power law distributions tend to describe systems of inter-acting elements, rather than just collections of variable ele-ments. Height is not a system—my height is independent of yours. My use of Wikipedia is not independent of yours, how-ever, as changes I make show up for you, and vice versa. This is one of the reasons we have a hard time thinking about sys-tems with power law distributions. We're used to being able to extract useful averages from small samples and to reason about the whole system based on those averages. When we encounter a system like Wikipedia where there is no represen-tative user, the habits of mind that come from thinking about averages are not merely useless, they're harmful. To under-stand the creation of something like a Wikipedia article, you can't look for a representative contributor, because none exists. Instead, you have to change your focus, to concentrate not on the individual users but on the behavior of the collective.

The power law also helps explain the difference between the many small but tightly integrated clusters of friends using weblogs and the handful of the most famous and best-trafficked weblogs. The pressures are reflected in Figure 5-2, which shows the relationship between a power law distribu-tion and the kinds of communication patterns that can be supported.

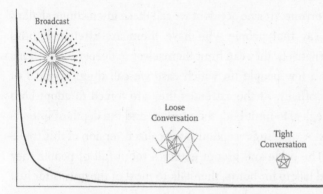

**Figure 5-2:** The relationship between audience size and conversational pattern. The curved line represents the power-law distribution of weblogs ranked by audience size. Weblogs at the left-hand side of the graph have so many readers that they are limited to the broadcast pattern, because you can't interact with millions of readers. As size of readership falls, loose conversation becomes possible, because the audiences are smaller. The long tail of weblogs, with just a few readers each, can support tight conversation, where every reader is also a writer and vice versa.

As is normal in a power law distribution, most writers have few readers. Such readers and writers can all pay similar amounts of attention to one another, forming relatively tight conversational clusters. (This is the pattern of small groups of friends using weblog or social networking tools, described in the last chapter.) As the audience grows larger, into the hundreds, the tight pattern of "everyone connected to everyone" becomes impossible to support—conversation is still possible, but it is in a community that is much more loosely woven. And with thousands of people paying attention, much less millions, fame starts to kick in. Once writers start getting more attention than they can return, they are forced into a width-versus-depth tradeoff. They can spend less time talking to everyone. (It's no accident we call these interactions shallow and say that people who have them are stretched thin.) Alternatively, they can limit themselves to deeper interactions with a few people (in which case we call them cliquish or standoffish). At the extremes they are forced to adopt both strategies, to limit both the number and the depth of interactions. A wedding reception is a localized version of this trade-off. The bride and groom gather a room full of people they could talk to for hours, then talk to most of the guests for just a few minutes each so as not to be rude.

## Why Would Anyone Bother?

Coase's logic in "The Nature of the Firm" suggests that in organizing any group, the choice is between management and chaos; he assumes that it's very difficult to create an unmanaged but nonchaotic group. But lack of managerial direction makes it easier for the casual contributor to add something of value; in economic terms, an open social system like Wikipedia dramatically reduces both managerial overhead and disincentives to participation. Even understanding how a wiki page comes into being does nothing to answer the hardest question of all: Why would anyone contribute to a wiki in the first place? The answer may be easiest to illustrate with a personal example.

I recently came across a Wikipedia entry for Koch snowflake, one of a curious bestiary of mathematical shapes called fractals (shapes that have the same pattern at many scales, like a fern leaf). The article had an animated image showing the snowflake in various stages of construction, accompanied by the following text:

> A Koch snowflake is the limit of an infinite construction that starts with a triangle and recursively replaces each line segment with a series of four line segments that form a triangular "bump." Each time new triangles are added (an iteration), the perimeter of this shape grows by a factor of 4/3 and thus diverges to infinity with the number of iterations. The length of the Koch snowflake's boundary is therefore infinite, while its area remains finite.

This description is accurate but a little awkward. I rewrote it to read:

> To create a Koch snowflake, start with an equilateral triangle and replace the middle third of every line segment with a pair of line segments that form an equilateral "bump." Then perform the same replacement on every line segment of the resulting shape, *ad infitum*. With every iteration, the perimeter of this shape grows by 4/3rds. The Koch snowflake is the result of an infinite number of these iterations, and has an infinite length, while its area remains finite.

This rewrite describes the same shape but in a way that is a little easier to grasp.

Why did I do it? Nothing in my daily life has anything to do with fractals, and besides, I was improving the article not for me but for subsequent readers. Psychological introspection is always a tricky business, but I know of at least three reasons I rewrote that description. The first was a chance to exercise some unused mental capacities—I studied fractals in a college physics course in the 1980s and was pleased to remember enough about the Koch snowflake to be able to say something useful about it, however modest.

The second reason was vanity—the "Kilroy was here" pleasure of changing something in the world, just to see my imprint on it. Making a mark on the world is a common human desire. In response to mass-produced technology with no user-serviceable parts inside, we install ringtones and screensavers, as a way to be able to change something about our inflexible tools. Wikipedia lets users make a far more meaningful contribution than deciding whether your phone should ring with the *1812 Overture* or "Holla Back Girl."

This desire to make a meaningful contribution where we can is part of what drives Wikipedia's spontaneous division of labor. You may have noticed that I accidentally introduced a mistake in my edit, writing *"ad infitum"* when I should have written *"ad infinitum."* I missed this at the time I wrote the entry, but the other users didn't; shortly after I posted my change, someone went in and fixed the spelling. My mistake had been fixed, my improvement improved. To propose my edit, I only had to know a bit about the Koch snowflake; there are many more people like me than there are mathematicians who understand the Snowflake in all its complexity. Similarly, fixing my typo required no knowledge of the subject at all; as a result, the number of potential readers who could fix my mistake was larger still, and because the fix was so simple, they did not need to have the same motivations I did. (If you noticed that error as printed here and were annoyed by it, consider whether that would have been enough to get you to fix it if you could.) It's obvious how Wikipedia takes advantage of different kinds of knowledge—someone who knows about World War II tank battles and someone who knows the rules of Texas hold'em are going to contribute to different articles. What's less obvious is how it takes advantage of skills other than knowledge. Rewriting a sentence to express the same thought more readably is a different skill from finding and fixing spelling errors, and both of those differ from knowing the rules of poker, but all those skills are put to good use by Wikipedia.

The third motivation was the desire to do a good thing. This motivation, of all of them, is both the most surprising and the most obvious. We know that nonfinancial motivations are everywhere. Encyclopedias used to be the kind of thing that appeared only when people paid for them, yet Wikipedia requires no fees from its users, nor payments to its contributors. The genius of wikis, and the coming change in group effort in general, is in part predicated on the ability to make nonfinancial motivations add up to something of global significance.

Yochai Benkler, a legal scholar and network theorist and author of *The Wealth of Networks*, calls nonmarket creation of group value "commons-based peer production" and draws attention to the ways people are happy to cooperate without needing financial reward. Wikipedia is peer production par excellence, set up to allow anyone who wants to edit an article to do so, for any and all reasons except getting paid.

There's an increasing amount of evidence, in fact, that

specific parts of our brain are given over to making economically irrational but socially useful calculations. In one well-known experiment, called the Ultimatum game, two people divide ten dollars between them. The first person is given the money and can then divide it between the two of them in any way he likes; the only freedom the second person has is to take or leave the deal for both of them. Pure economic rationality would suggest that the second person would accept any split of the money, down to a $9.99-to-$.01 division, because taking even a penny would make him better off than before. In practice, though, the recipient would refuse to accept a division that was seen as too unequal (less than a $7-to-$3 split, in practice) even though this meant that neither person received any cash at all. Contrary to classical economic theory, in other words, we have a willingness to punish those who are treating us unfairly, even at personal cost, or, to put it another way, a preference for fairness that is more emotional than rational. This in turn suggests that relying on nonfinancial motivations may actually make systems more tolerant of variable participation.

We also have practical evidence that when a perceived bargain changes, previously contented volunteers will defect. America Online built its business as a user-friendly entry point into digital networks, and much of its friendliness came directly from AOL's users, many of whom loved the service so much that they worked as volunteer guides. After AOL's stock price rose into the stratosphere, however, a number of those guides banded together to file a class-action suit, claiming AOL had unfairly profited from their work. Nothing had changed about the job they were being asked to do; everything changed about the financial context they did it in, and that was enough to poison their goodwill. (Though the case is still pending, AOL has dropped the volunteer guide program.)

## Social Prosthetics

The question every working wiki asks of its users is "Who cares?" Who cares that an article on asphalt exists? Cdani does. Who cares that it include photos? SCEhardt does. Who cares that the Koch snowflake description be clear? I do. Wikis reward those who invest in improving them. This explains why both experts and amateurs are willing to contribute—the structure of participation is not tied to extrinsic rewards, so people capable of adding to the technical explanation of complex mathematical shapes end up working alongside people who only know enough to be able to proofread descriptions of same. This reward, and the loyalty it creates, help explain one of the most complex questions about Wikipedia's continued success: How does it survive both disagreement and vandalism? Openness, division of labor, and the multiple motivations of its users drive its rising average quality, but none of those things explain why articles on contentious subjects aren't damaged by editing wars among rival factions, or simply destroyed by vandals, who can delete an entire article with the click of a button. Why don't these sorts of things happen? Or to ask the same question in the language of economics: Why doesn't Wikipedia suffer from the Tragedy of the Commons? Why haven't free riders and even vandals destroyed it?

The wiki format is another version of publish-then-filter;

coercion is applied after the fact rather than before. All edits are provisional, so any subsequent reader can decide that a change to an article is unacceptable, to be further edited or to be deleted. This capability is universal; any edit or deletion can be further edited or undone ("reverted"), changes that are themselves then held up for still more scrutiny, ad infinitum. Every change to a Wikipedia article is best thought of as a proposed edit; it shows up the minute it is made, but it is still subject to future review and revision. (I checked back on the Koch snowflake article later and was pleased to see my changes had survived such review.) In the case of obvious vandalism, the review process happens astonishingly quickly. Martin Wattenberg and Fernanda Viegas, researchers at IBM who study Wikipedia, have documented a number of contentious articles on subjects like abortion and Islam where complete deletions of the articles' contents have been restored in less than two minutes.

Like everything described in this book, a wiki is a hybrid of tool and community. Wikipedia, and all wikis, grow if enough people care about them, and they die if they don't. This last function is part of any working wiki, but it isn't part of the wiki software, it's part of the community that uses the software. If even only a few people care about a wiki, it becomes harder to harm it than to heal it. (Imagine a world where it was easier to clean graffiti off a wall than to put it there in the first place.) When a vandalized page reappears as if nothing has happened, it creates the opposite of the "Kilroy was here" feeling of a successful edit—nothing is more frustrating to a vandal than investing energy to change something and then have that effort disappear in seconds. Evidence that enough people care about an article, and that they have both the will and the tools to quickly defend it, has proven enough to demoralize most vandals.

As with every fusion of group and tool, this defense against vandalism is the result not of a novel technology alone but of a novel technology combined with a novel social strategy. Wikis provide ways for groups to work together, and to defend the output of that work, but these capabilities are available only when most of the participants are committed to those outcomes. When they are not, creating a wiki can be an exercise in futility, if not an outright disaster. One notable example was the *Los Angeles Times* "Wikitorial" effort, in which the content of the paper's editorial pages was made available to the public. The *Times* announced the experiment in a bid to drive users there, and drive them they did. A group of passionate and committed users quickly arrived and set about destroying the experiment, vandalizing the posted editorials with off-topic content and porn. The Wikitorial had been up for less than forty-eight hours when a *Times* staffer was told to simply pull the plug. The problem the *Times* suffered from was simple: no one cared enough about the contents of the Wikitorial to defend it, much less improve it. An editorial is meant to be a timely utterance of a single opinionated voice—the opposite of the characteristics that make for good wiki content. A wiki augments community rather than replacing it; in the absence of a functioning community, a wiki will suffer from the Tragedy of the Commons, as the Wikitorial did, as individuals use it as an attention-getting platform, and there is no community to defend it.

One of the extreme defensive strategies for Wikipedia is

the ability to lock a page, preventing all but a few of the most committed Wikipedians from editing it until passions have cooled. (Pages can be locked in the face of sustained vandalism as well, but at any given time less than half a percent of pages are locked.) In addition there have been crises of validity: in 2005 the journalist John Seigenthaler, Sr., discovered that he had a biography on Wikipedia and that it contained scurrilous and false accusations about his involvement in the Kennedy assassinations. The entry was then fixed, but by that time the false material had been in place for half a year, so much of the damage had been done. Then in 2006 a longtime Wikipedian going by essjay claimed, among other things, that he had a doctorate in theology and worked as a tenured professor at a private university. In fact, he had dropped out of a community college and had no degree or academic job of any sort. Both of these events demonstrated weaknesses in Wikipedia's methods, and in the aftermath of each one the Wikimedia foundation instituted new rules, including special proposals for handling biographies of the living, as well as restricting the ability of unregistered users to create articles from scratch.

The locking of pages and the restrictions put in place after the Seigenthaler and essjay affairs contrast with Wikipedia's general goal of openness. The Wikipedians are acutely aware of this conflict and as a result the design philosophy hews to Cunningham's original work: let the community do as much as they possibly can, but where they can't do the work on their own, add technological fixes. Wikipedia is predicated on openness not as a theoretical way of working but as a practical one. This pragmatism often comes as a shock to people who hold up Wikipedia as a beacon of pure openness, but the curious fact is that many of Wikipedia's most vociferous boosters actually don't know much about its inner workings and want to regard it, wrongly, as an experiment in communal anarchy. The people most enamored of describing Wikipedia as the product of a free-form hive mind don't understand how Wikipedia actually works. It is the product not of collectivism but of unending argumentation. The articles grow not from harmonious thought but from constant scrutiny and emendation.

The idea behind Nupedia was that it should be possible to improve on traditional encyclopedias by keeping the process but dropping the commercial aspect. This turned out to be a bad idea, because much of the process for creating a traditional encyclopedia has less do with encyclopedias than with institutional imperatives. Once you dispense with the institutional dilemma, as Wikipedia does, it is possible to dispense with much institutional process as well. Wikipedia invites us to do the following disorienting math: a chaotic process, with unpredictable and wildly uneven contributions, made by nonexpert contributors acting out of variable motivations, is creating a global resource of tremendous daily value. A commercial producer of encyclopedias has to be efficient about finding and fixing mistakes, since things like process and deadlines and salaries are involved. Wikipedia, with none of those things, does not have to be efficient—it merely has to be effective. If enough people see an article, the chance that an error will be caught and fixed improves with time. Because Wikipedia is a process, not a product, it replaces guarantees offered by institutions with probabilities supported by process: if enough people care enough about an article to read it, then enough people will care enough to improve it, and over time this will lead to a large enough body of good enough work to begin to take both availability and quality of articles for granted, and to integrate Wikipedia into daily use by millions.

## Love as a Renewable Building Material

The Ise Shrine, a Shinto shrine in Ise, Japan, has occupied its current site for over thirteen hundred years. Despite its advanced age, however, UNESCO, the UN cultural agency, refused to list the shrine in its list of historic places. Why? Because the shrine is made out of wood, never a material prized for millennium-scale structural integrity, and so it can't be thirteen hundred years old. The Imbe priests who keep the shrine know that too, but they have a solution. They periodically tear the shrine to the ground, and then, using wood cut from the same forest that the original was built from, they rebuild the shrine to the same plan, on an adjacent spot. They do this every couple of decades and have done it sixty-one times in a row. (The next rebuilding will be in 2013.) Because the purpose of the shrine is in part to delineate the difference between sacred and ordinary space, from their point of view they have a thirteen-hundred-year old shrine, built out of renewable materials. This argument didn't wash with UNESCO; the places they list enjoy the solidity of edifice, not of process. A wrecked castle that has stood unused for five hundred years makes the cut; a shrine that is rebuilt once a generation for a thousand years doesn't.

Wikipedia is a Shinto shrine; it exists not as an edifice but as an act of love. Like the Ise Shrine, Wikipedia exists because enough people love it and, more important, love one another in its context. This does not mean that the people constructing it always agree, but loving someone doesn't preclude arguing with them (as your own experience will doubtless confirm). What love does for Wikipedia is provide the motivation both for improvement and for defense. If the company that makes *Encyclopaedia Britannica* were to go out of business tomorrow, its core product would slowly decay as new knowledge was accrued without being reflected in subsequent editions. This concept, sometimes called the half-life of knowledge (in metaphorical comparison to radioactive decay), would render *Britannica* obsolete as the years wore on. If the people who love Wikipedia all lost interest at the same time, on the other hand, it would vanish almost instantly. The vandals and special interest groups who are constantly fighting to alter articles fail only because people care about Wikipedia, both article by article and as a whole, and because Wikipedia as a tool provides them with the weapons to fight those groups. Those weapons are taken up only by people who are willing to fight. Were that willingness to fade, the most contentious articles on Wikipedia, the articles on abortion and Islam and evolution, would be gone within hours, and it's unlikely that the whole enterprise would survive a week.

We don't often talk about love when trying to describe the public world, because love seems too squishy and too private. What has happened, though, and what is still happening in our historical moment, is that love has become a lot less squishy and a lot less private. Love has a half-life too, as well as a radius, and we're used to both of those being small. We can affect the people we love, but the longevity and social distance of love are both constrained. Or were constrained—now we can do things for strangers who do things for us, at a low enough cost to make that kind of behavior attractive, and those effects can last well beyond our original contribution. Our social tools are turning love into a renewable building material. When people care enough, they can come together and accomplish things of a scope and longevity that were previously impossible; they can do big things for love.

# CHAPTER 6

# COLLECTIVE ACTION AND
# INSTITUTIONAL CHALLENGES

{ *Collective action, where a group acts as a whole, is even
more complex than collaborative production, but here
again new tools give life to new forms of action. This in turn
challenges existing institutions, by eroding the institutional
monopoly on large-scale coordination.* }

A t the beginning of 2002 the *Boston Globe* published a two-
part series detailing the history of Father John Geoghan, a
Catholic priest. Geoghan had worked at various parishes in the
Boston archdiocese since the 1960s and during that time had
fondled or raped more than a hundred boys. The *Globe* reporters reviewed documents the church had been forced to submit
for the Geoghan case, which revealed accusations against him
dating back to the 1960s. The church's response had been to
move him from parish to parish, sometimes with therapeutic
stints in between. If these treatments helped, their effect was
temporary; the abuse continued over a thirty-five-year period.

Cardinal Bernard F. Law, then archbishop of the Boston Diocese, had known about Geoghan's problems as early as 1984 but continued the pattern of periodic relocation; Geoghan was not removed from the priesthood until 1998.

The *Globe* story ignited a firestorm of controversy among a shocked Catholic laity. A few weeks after the *Globe* articles appeared, a physician named James Muller hosted a gathering for those determined to channel their shock and anger into some sort of productive change. The group met for the first time on a freezing January night in the basement of a church in the Boston suburb of Wellesley; thirty people showed up. The impetus for the original meeting was to react somehow to the horror of the priestly abuse and the bishops' failures in handling it, but in talking it over that night, the assembled members of the laity decided to pursue a more coordinated strategy of activism, and a group called Voice of the Faithful (VOTF) was born.

There was nothing terribly new about the founding of VOTF. Small groups of concerned citizens have been meeting in church basements and public libraries for a long time, and Muller, the founder, was no stranger to organizing, having helped found International Physicians for the Prevention of Nuclear War in the 1980s. VOTF's growth, though, was new. After just two months in existence VOTF meetings were attracting large crowds. Muller later wrote of a March meeting, "I was unable to park within four blocks as more than five hundred people overflowed our small meeting rooms." By the summer of 2002, when VOTF held its first convention, it had 25,000 members, four thousand of whom came to Boston to attend. The organization's growth was also international:

VOTF had members in more than twenty countries in its first year of existence. It's difficult to convey what a torrid pace of growth this is—to go from thirty people in a church basement to 25,000 is almost a thousandfold increase. To do so in half a year meant doubling the size of the organization every couple of weeks. Groups don't grow that fast for any sustained period, or rather they didn't until the barriers to that kind of growth were removed.

VOTF adopted a slogan—"Keep the Faith, Change the Church"—that made it clear that they were not going to be content with expressing simple outrage. They wanted structural change. The boldness of their demands left the church unsure how to react. By April, when it became clear that VOTF was still gaining members, Cardinal Law asked Bishop Walter J. Edyvean, one of his deputies, to arrange a meeting. It did not go well. During that meeting Edyvean said he and Law had "issues" with VOTF and admitted under repeated questioning to having tried to block VOTF meetings from taking place on church property. VOTF members were shocked by the completeness of the opposition, but at the end of the meeting, in the spirit of comity, they agreed to issue a joint statement saying they had had a productive meeting. That was the high-water mark of cordiality between the two groups.

After the meeting came public pronouncements from the church. Edyvean declared that no parish was to allow VOTF to meet on church property, and that any valid lay organization had to operate "exclusively within the parish where it has been established" and be "presided over by the pastor of that parish." The church's position was that even the laity were contained by the church's hierarchical structure; there was to be no lay conversation across parish lines whatsoever. As the year wound down and the scandal did not, there were increasingly broad and public calls for Law to step down. Law refused to consider either engagement with VOTF or resignation from the archdiocese, but his objections proved ineffective on both counts. On November 26, after almost a year of steadfast refusal, Law had his first meeting with VOTF. The meeting was cordial but inconclusive; both sides again said they hoped for a constructive outcome to the dialogue. What no one but perhaps Law knew was that this first meeting would also be his last. He traveled to Rome to offer his resignation to Pope John Paul II, who accepted it on December 13.

## Why 2002? What Changed?

It is understandable that Law resigned, and that the church began to take steps, if halting ones, to publicly reform itself. The bishops had not merely sheltered abusive priests; they had done so in a way that endangered still more of their parishioners. What is not as clear is "Why then?" Given that much of the reported abuse was in the 1960s and 1970s, why was 2002 the year that scandal rocked the church? There are three obvious answers: the abuse had become too extensive to ignore, the existence of abusive priests had become public in a court of law, and a particularly horrific case was getting serious media coverage. The interesting thing about those answers is that none of them really hold water.

In May 1992, a decade before the Geoghan case, another Catholic priest, the Reverend James R. Porter, was accused of sexual abuse of children in three different Boston parishes. (Ninety-nine people eventually came forward with accusations against Porter.) Like Geoghan, Porter had been quietly moved from one parish to the next. The *Boston Globe* covered that case as well, and Bernard Law, then the local bishop, criticized the coverage as unfair and, from the pulpit, called for divine retribution saying, "By all means we call down God's power on the media, particularly the *Globe*." (Law's prayer was not answered: the *Globe* went on to publish more than fifty stories and editorials on the Porter case and on priestly abuse in general in that year.)

In 1992 the Porter case offered the same raw material for outraged reaction, in the same diocese, with many of the same actors as the Geoghan case a decade later: horrific abuse had become public in a court of law, accompanied by a flurry of media coverage. Nonetheless in 1992 the outrage dissipated with little change in the church's behavior in Massachusetts or nationally, no official reaction from the Vatican, no coordinated calls by the laity for Law's resignation, and no resignation. Law may have been sanguine about refusing to resign in 2002 because he'd been through an eerily similar set of events a decade earlier, and the crisis had abated.

The church's strategy for both the Porter and the Geoghan cases was to treat the matter as an internal affair; an individual parish might be scandalized for a short period, but with no victims talking and little public record, the sense of scandal could neither spread nor last. In this way the church could avoid major and synchronized outrage. This was a strategy not for ending the abuse but for managing the fallout. Something happened between the Porter case in 1992 and the Geoghan case in 2002, something that destroyed the effectiveness of the church's strategy. In 1992 Law relied on two facts: ordinary Catholics couldn't easily share information about the scandal with one another, and they couldn't readily coordinate their response. By 2002 those two facts had ceased to be facts.

## New Forms of Sharing Take Hold

The old limits on sharing information were the first thing to change. This change is essential, and not just for the abuse scandal. The impulse to share important information is a basic one, but its manifestations have often been clunky. Consider the *Globe*'s story on Porter in 1992—even the seemingly minor difficulties of clipping and mailing a newspaper article were significant enough to greatly limit the frequency of that kind of forwarding. The cumulative effects of those difficulties were even stronger—to clip an article and share it with a group, you would have to copy it first, adding a step and thus reducing the attractiveness of sending it in the first place. Similarly, the recipient of a mailed clipping can't both forward it and keep it without reincurring all the difficulties of the original sender. As a result of these difficulties, the readership for any given newspaper story was a subset of the readership for the paper generally.

By 2002 those difficulties had vanished. The *Globe* was online in the form of Boston.com, its readers had e-mail, and some even had weblogs. The act of forwarding a story to friends and colleagues had gone from tedious to all but effortless. Even more important, forwarding the story to a group was as easy as forwarding it to an individual, and any of the recipients could forward it to others as easily as the original sender had done. Whereas a newspaper relies on an asymmetry of production versus consumption—readers don't own printing presses—any recipient of e-mail can also be a sender, by definition. Now the readership for a particular story can be larger than the paper's general audience, as with the Geoghan case or the *Times-Picayune*'s coverage of Hurricane Katrina.

The social urge to share information isn't new. Prior to e-mail and weblogs, we clipped articles and published family newsletters. Recalling these older behaviors, it's tempting to conclude that our new tools are merely improvements on existing behaviors; this view is both right and wrong. The improvement is there, but it is an improvement so profound that it creates new effects. Philosophers sometimes make a distinction between a difference in degree (more of the same) and a difference in kind (something new). What we are witnessing today is a difference in the degree of sharing so large it becomes a difference in kind. Prior to e-mail and the Web, we could still forward and comment on the news of the day, but the process was beset by numerous small difficulties. The economic effects of even seemingly insignificant hurdles are counterintuitive but remarkable: even the minimal hassle involved in sending a newspaper clipping to a group (xeroxing the article, finding envelopes and stamps, writing addresses) widens the gap between intention and action.

In 1992 the *Globe* wasn't global, and the Porter story stayed in Boston. In 2002 the *Globe* didn't need to spread the Geoghan story to the world's Catholics; the world's Catholics were capable of doing that themselves. The effect of such reader redistribution was so significant that the New York Times Company, the parent of the *Boston Globe*, specifically referred to the Geoghan coverage in an investor report, noting that the story's popularity was a significant factor in reaching new readers and in raising the number of readers overall.

The year 2002 also saw strong growth for weblogs, and a variety of bloggers were talking about the story, thus creating both a clearinghouse for new stories and a long-lived archive of past stories. The motivations of these bloggers were quite varied. While they expressed universal horror at Geoghan's actions, many of the subsequent discussions were very skeptical of lay involvement in matters involving priests. Voice of the Faithful, in particular, received strong criticism as a group of radicals, which in turn kicked off heated arguments. Even the most vigorous attacks on the church's critics, though, had the effect of increasing awareness of the scandal and of the existence of organizations like VOTF. None of these were effects the church had ever faced before. Every time someone criticized Law, or pointed to the *Globe* story, or even condemned the church's critics, the church's ability to wait out the scandal eroded a little more.

The low cost of aggregating information also allowed the formalization of sharing among people tracking priestly abuse. BishopAccountability.org, launched a year after the Geoghan case, collated accusations of abuse, giving a permanent home to what in the past would have been evanescent coverage. David Clohessy, the director of Survivors Network of those Abused by Priests (SNAP), credits the ability to collect and share information with the change in public perception: "What technology did here was help expose the lie in the two greatest

PR defenses of this kind of abuse: 'This is an aberration' and 'We didn't know.' When you can send a reporter twenty links to nearly identical stories, then that reporter obviously approaches his or her own bishop with greater skepticism and much more vigor."

## Rapid and Simple Group Formation

As important as improved information sharing is, it's only part of this story. Easier and wider dissemination of information changes group awareness, but even that would have had a limited effect without a change in collective action as well. Had VOTF been founded in 1992, the gap between hearing about it and deciding to join would have presented a set of small hurdles: How would you locate the organization? How would you contact it? If you requested literature, how long would it take to arrive, and by the time it got there, would you still be in the mood? No one of these barriers to action is insurmountable, but together they subject the desire to act to the death of a thousand cuts.

Because of the delays and costs involved, going from a couple dozen people in a basement to a large and global organization in six months is inconceivable without social tools like websites for membership and e-mail for communication. A survey of VOTF members conducted by the Catholic University of America in 2004 noted with some puzzlement that many members of VOTF were not members of a regional VOTF affiliate; their connection was with the institution directly. As the report noted, "In a sense, the Internet becomes a kind of affiliate for many." The same survey reported that a majority of members had attended their first VOTF event alone; where most membership organizations grow because someone is invited by a friend or neighbor, VOTF grew when people were looking for information online. This change also affected existing organizations; SNAP had nine chapters after being in existence twelve years; in 2002 it added thirty-five more, a yearly rate of growth fifty times the previous norm.

VOTF has become a powerful force, all while remaining loosely (and largely electronically) coordinated. As John Moynihan of VOTF puts it, "Between 2002 and now, we've changed from an affiliate model to an internet model." VOTF is now at a crossroads—five years after the precipitating crisis, it is facing a budget shortfall. This is a common event among organizations that grow quickly; they factor in the rapid growth, and when it slows, as it inevitably must past a certain size, they feel the pinch. With many more possible groups competing for the average individual's time, the speed with which a group can come unglued has also increased. Partly in response to the slowing of growth, VOTF has adopted a more direct opposition to Catholic doctrine and is preparing a campaign to argue against enforced celibacy of Catholic priests, on the grounds that this requirement has contributed to the problem of abusive priests. This oppositional stance will make it even more of a lighting rod for its critics, but this opposition may in turn rally some of VOTF's members more strongly to its cause. Whatever happens, though, VOTF is heading in the direction of even stronger collective action; by committing itself to a more explicitly contentious plan of action, it will provide evidence for how far the "internet model" for assembling a group can go in getting that group to act together in the face of significant opposition.

## Removing Obstacles to Collective Action

Technology didn't cause the abuse scandal that began in 2002. The scandal was caused by the actions of the church, and many factors affected the severity of reaction in 2002, including the exposure of more of the church's internal documents and the effectiveness of the *Globe*'s coverage. That combination was going to lead to substantial reaction in any case. What technology did do was alter the spread, force, and especially duration of that reaction, by removing two old obstacles—locality of information, and barriers to group reaction.

The Catholic Church is one of the oldest continually operating institutions in the world, and has had a hierarchical system of management for well over a thousand years, long enough to have lived through radical technological change before. Gutenberg's improvement of the printing press with movable type helped catalyze the Protestant Reformation in Europe in the 1500s. Then as now, a power previously vested in the Catholic hierarchy has become broadly accessible. When the invention was the printing press, the result was direct access to the text of the Bible in languages other than Latin. Today, with social tools, it is organizational participation by the laity. Though it is much too early to tell if this change in communications technology, and the attendant challenge to the church, will be as significant in its ramifications, the basic struggle is the same, and it isn't just vested in VOTF. Many organizations with lay membership are benefiting from novel forms of collaborative tools.

For most of church history, the priestly hierarchy *was* the church; the laity would gather in the parishes, but all of the power and all of the decision-making was with the priesthood—even after the 1960s, when the famous Second Vatican Council declared that the Church was composed of both the priests and the parishioners. Whether by design or accident, Vatican II, as it came to be called, was more of a feel-good nostrum than an actual recipe for change—it was fine to suggest that the laity somehow constituted the body of the church, but without a mechanism for allowing Catholics to make their feelings known, the practical effect on the hierarchy was minimal. Over the centuries the Catholic Church has been buffeted by incredible institutional pressures, but in all that time every real push for change has come from within the priesthood, from Martin Luther's 95 *Theses* in the 1500s to the Liberation Theology of Central and South America in the 1980s. No significant challenge to the hierarchy has ever come directly from the laity—until now. The reaction of the Catholic laity to the abuse scandal is showing us one way in which Vatican II might be implemented, how a collection of individuals previously obstructed from sharing information and opinions across parish lines can have a lasting effect on the church by working together as a group.

The Catholic Church is not the only organization affected by this. In 2007 several conservative parishes of the Episcopalian Church in Virginia voted to break off from the American church in protest over the ordination of an openly gay bishop, Gene Robinson. Instead of forming their own breakaway church, though, the parishes joined the Nigerian church, whose bishop, Peter Akinola, is deeply antagonistic to homosexuals' involvement in the church in any form. The idea that a church in Fairfax, Virginia, could simply declare itself part of another diocese on a different continent upends centuries of tradition. As with Edyvean's demands that parishioners not organize across parish lines, Anglican bishops are not to control churches outside the geographic boundaries of their diocese. What the Virginia diocese has done is not to relocate but to de-locate. By announcing that Virginia churches are part of the Nigerian diocese, in contravention of all geographic sense, the Virginians are doing more than voting their conscience on the issue of acceptance of gays; they are challenging geography as an organizing principle for the church. In a world where group action means gathering face-to-face, people who need to act as a group should, ideally, be physically near one another. Now that we have ridiculously easy group-forming, however, that stricture is relaxed, and the result is that organizations that assume geography as a core organizing principle, even ones that have been operating that way for centuries, are now facing challenges to that previously bedrock principle.

Before social tools were widely available, the church didn't have to forbid the laity to organize across parish lines—it wasn't a possibility in the first place. Edyvean's demand that ordinary Catholics not cross parish lines was an attempt to replace by doctrine what was no longer enforced by physical obstacles. What Edyvean didn't understand was that the lack of previous historical challenges from the laity was an example not of forbearance, but of inability. Vatican II's promise of engagement was, until very recently, an empty one, but no longer. Some new and stable arrangement will eventually be found, as it was after Martin Luther, but whatever it is, the one option that it won't include is a return to the days of a subdivided and disorganized laity.

It is a curiosity of technology that it creates new characteristics in old institutions. Prior to the spread of movable type, scribes didn't write slowly; they wrote at ordinary speed, which is to say that in the absence of a comparable alternative, the speed of a man writing was the norm for all publishing. After movable type came in, scribes started to write slowly, even though their speed hadn't changed; it was simply that they were being compared to something much faster. Similarly, prior to this decade, the Catholic Church was not inimical to improvised global organization of its parishioners, because it simply wasn't an option; even for a group that believes in miracles, that kind of thing was obviously outside the realm of possibility. Now that it is an option, the church has to react, and that reaction, forced by the presence of groups using social tools, is to fight against something that ten years ago wasn't even an abstract possibility.

## Ordinary Tools, Extraordinary Effects

Once something becomes ordinary, it's hard to remember what life was like without it, but it's worth remembering that before e-mail we had few tools for group communication, none of them very good. What does e-mail have going for it that the other attempts at many-to-many communication didn't? Cost, for one. E-mail across the ocean costs no more than e-mail around the block, and e-mail to ten people costs no more than e-mail to one. With e-mail, having a large, long-lived, and geographically widespread conversation entails no expenses. E-mail delivery is almost instant, unlike ordinary mail, but doesn't require the sender and the receiver to be synchronized with one another, unlike the phone. This asynchrony reduces transaction costs for group communication in the same way that the economic model of e-mail reduces the dollar costs. These advantages help account for the incredible success of e-mail as a medium for group conversation, relative to all previous attempts.

These features aren't really advantages of e-mail per se. The earliest e-mail programs, written in the 1970s, were incredibly simple tools, yet the advantages of cost and asynchrony were already there. They were built into the network on which e-mail is built: the internet. The internet is the first big communications network to make group communication a native part of its repertoire. The basic logic of the internet, called "end-to-end communication," says that the internet itself is just a vehicle for moving information back and forth—it's up to the computers sending and receiving information to make sense of it. While the telephone network was engineered for transmission of voice (and the phone company fought bitter legal battles to keep it from being used for any other purpose), the internet does not know what it is being used for. This fact has many ramifications, but two of the most important ones are vanishingly cheap many-to-many communications, and the flexibility that allows people to design and try new communications tools without having to ask anyone for permission. The most important of these experiments has been the Web. Begun as a research effort in the early 1990s by Sir Tim Berners-Lee (knighted, in fact, for that invention), the Web became a core part of modern life as quickly as it did precisely because it is such a flexible environment for letting people try new things.

The communications tools broadly adopted in the last decade are the first to fit human social networks well, and because they are easily modifiable, they can be made to fit better over time. Rather than limiting our communications to one-to-one and one-to-many tools, which have always been a bad fit to social life, we now have many-to-many tools that support and accelerate cooperation and action.

And the possibilities for global organization enabled by those tools continue to grow. A more recent challenge to the Catholic Church has come on the heels of the Boston revelations: in 2006 the BBC aired a documentary, *Sex Crimes and the Vatican,* on the church's handling of sexual abuse cases by priests. Shortly after it aired, the Italian TV channel RAI acquired rights to show the video, but members of the ruling party, the church, and managers at RAI objected. Believing that the documentary deserved to be better known in Italy, a group of bloggers operating at Bispensiero.it took matters into their own hands. They subtitled the forty-minute documentary in Italian themselves, then posted it to a video hosting site, where it has been viewed more than a million times. *Avvenire,* the newspaper of Italy's Conference of Bishops, attacked the video as slander, but once it became available in Italy on the Web, the matter was effectively decided: RAI aired its version in early June, several days after Bispensiero forced its hand. The first significant challenge to the church from the newly organized laity wasn't an anomaly—it was the beginning of an era. VOTF's ability to use the *Boston Globe* article as the rallying point for group action was the first of many such events the church will face in coming years.

One way to think about the change in the ability of groups to form and act is to use an analogy with the spread of disease. The classic model for the spread of disease looks at three variables—likelihood of infection, likelihood of contact between any two people, and overall size of population. If any of those variables increases, the overall spread of the disease increases as well. This model also applies well to the spread of gossip and other word-of-mouth opinion. What happened in the Boston archdiocese between 1992 and 2002 is that both the size of the audience and the ease of contact increased dramatically. As a result, the spread of information and its value as a coordinating force increased dramatically as well. (Much of the advertising world, in fact, has spent the last several years pursuing "viral marketing" on exactly this analogy.) What the rise of new and newly powerful lay organizations shows us is that in the right cases people are willing and even eager to come together and affect the world. Motivation, energy, and talent for action are all present in those sorts of groups—what was not present, until recently, was the ability to coordinate easily.

Seen in that light, social tools don't create collective action—they merely remove the obstacles to it. Those obstacles have been so significant and pervasive, however, that as they are being removed, the world is becoming a different place. This is why many of the significant changes are based not on the fanciest, newest bits of technology but on simple, easy-to-use tools like e-mail, mobile phones, and websites, because those are the tools most people have access to and, critically, are comfortable using in their daily lives. Revolution doesn't happen when society adopts new technologies—it happens when society adopts new behaviors.

# CHAPTER 7

# FASTER AND FASTER

*As more people adopt simple social tools, and as those tools allow increasingly rapid communication, the speed of group action also increases, and just as more is different, faster is different.*

Collective action is different from individual action, both harder to get going and, once going, harder to stop. As Judge Richard Posner put it, "Conspiracies are punished separately from single-offender criminal acts, and often as severely even if the conspiracy fails to achieve its aim, because a group having some illegal purpose is more dangerous than an individual who has the same purpose." This is true not just of criminal intent. Groups are capable of exerting a different kind of force than are individuals, and when that force is turned against an existing institution, groups create a different kind of threat.

To understand the difference, consider the events of 1989 in the East German city of Leipzig. At the beginning of that year a handful of Leipzigers began protesting against the German Democratic Republic (GDR), often staging these protests during an existing event—a street music festival, a fair—that offered a way to get a mass of people together without arousing suspicion. At first the protests were small—in January five hundred people showed up, and the government arrested fifty of them. That didn't deter the protesters, however. As the year progressed, the protests became more regular, taking place every Monday. When each subsequent protest came and went, more bystanders realized that the government was doing nothing systematic to stop them. As a result, every subsequent Monday new participants joined in, which in turn emboldened still more citizens.

Early on the marches were too small for the government to stop without looking hysterical, and every week they grew only a little. From the government's point of view, a small march was too little to crack down on, and the following week, a slightly larger march was also too little to crack down on. Not until September did Erich Honecker instruct local governments to "nip these enemy activities in the bud" and "not allow a mass basis for them." By then it was too late; the protests had long since passed from bud to full flower. What Honecker could not have known was that the "mass basis" was measured not by the number of participants but by the number of people who understood that protest was not being punished. The historian Susanne Lohmann calls the Leipzig protests an "information cascade." Each of the citizens of Leipzig had some threshold at which they might join a protest. Every week the march happened without a crackdown offered additional evidence that the marches provided an outlet for their disaffection; each successful march diminished the fear felt by some additional part of the populace.

The military often talks about "shared awareness," which is the ability of many different people and groups to understand a situation, and to understand who else has the same understanding. If I see a fire break out, and I see that you see it as well, we may more easily coordinate our actions—you call 911, I grab a fire extinguisher—than if I have to call your attention to the fire, or if I am in some confusion about how you will react to a fire. Shared awareness allows otherwise uncoordinated groups to begin to work together more quickly and effectively.

This kind of social awareness has three levels: when everybody knows something, when everybody knows that everybody knows, and when everybody knows that everybody knows that everybody knows. Many people in the GDR figured out for themselves that the government was corrupt, and that life under that government was bad; this is the "everyone knows" condition. Over time many of those same people figured out that most of their friends, neighbors, and colleagues knew that as well—"everyone knows that everyone knows." At this point the sentiment was widespread but because no one was talking about what everyone knew, the state never had to respond in any formal way. Finally people in Leipzig could see others acting on the knowledge that the GDR was rotten—"everyone knows that everyone knows that everyone knows." This shared awareness is the step necessary for real public action: when the people in the streets of Leipzig knew the same thing as did the people watching from their windows.

By September 1989 this information had cascaded from a

small group to a large one, and the marches had grown to tens of thousands of people. In October the number grew to better than a hundred thousand. On the first Monday in November 400,000 people turned out in the streets of Leipzig. By the time the government realized its bluff was being called, no one in the army was willing to turn on so many citizens, and without a credible threat of deadly force to back it up, the East German government simply collapsed. The day after that first November protest the entire East German government resigned. Two days later the dismantling of the Berlin Wall began. The GDR had vanished.

The lesson for protesters after Leipzig was that they should protest in ways that the state was unlikely to interfere with, and to distribute evidence of their actions widely. If the state didn't react, the documentation would serve as evidence that the protesting was safe. If the state *did* react, then the documentation of the crackdown could be used to spur an international outcry. The lesson for repressive states was the opposite: don't let even small protests get started, as they can grow, and don't let any documentation get out. These two lessons set up a cat-and-mouse game between protesters and the protested institutions that continues to this day. As in everything that involves coordinated action, social tools have changed the balance of power in this game.

## Flash Mobs

Early one June evening in 2003, more than a hundred people arrived on the ninth floor of Macy's department store, where they proceeded to look at one particular large and very expensive rug. When the puzzled sales assistant asked if they needed help, the members of the group explained that they lived together in a commune, were shopping for a "love rug," and made all their decisions in a group. Then, ten minutes later, the crowd suddenly dispersed, heading in different directions with no obvious coordination.

The event was the first successful flash mob, a group that engages in seemingly spontaneous but actually synchronized behavior. The form was invented by Bill Wasik, an editor at *Harper's* magazine, as a kind of street performance, as well as an ironic commentary on the conformism of hipster culture. Wasik, working as the anonymous "Bill from New York," would e-mail instructions to a group of people, spelling out when and where they were to converge and describing the activity they were to engage in once there. Later flash crowds involved getting dozens of people to perch on a stone ledge in Central Park all making bird noises, a "Zombie walk" in San Francisco, and a silent dance party at London's Victoria Station. These mobs had some of the flavor of flagpole sitting—harmless but attention-getting fun. But as the novelist William Gibson noted about technology, the street finds its own uses for things, and after their flagpole-sitting phase, flash mobs entered the political sphere.

The first use of a flash mob for political expression appeared soon after the "love rug" mob. Howard Dean's U.S. presidential campaign proposed a flash mob in Seattle in September. (The invitation was published in Garry Trudeau's *Doonesbury* cartoon.) The next year protesters staged a flash mob against Russian prime minister Vladimir Putin in his home city of St. Petersburg, two weeks before the Russian presidential elections. About sixty youths turned up in Putin masks, wearing shirts with anti-Putin messages like "Vova go home!" (Vova is a nickname for Vladimir.) The use of flash mobs as a tool of political protest, though, has reached its zenith in Belarus.

Belarus is one of Europe's most repressive countries. A former member of the Union of Soviet Socialist Republics, it was cut loose after the collapse of European Communism during the 1990s. In the main, the former Soviet states embraced free markets and democratic process, but Belarus retained a state-run economy and acquired an autocratic president, Alexander Lukashenko, who was first elected in 1994 on a platform of eradicating corruption. In the intervening years Lukashenko has ruled over the country with increasingly unchecked power. When he ran for reelection to a third term in March 2006, he won nearly 85 percent of the vote, a result that European election observers said was rigged. In protest, more than ten thousand people turned out in Minsk's Oktyabrskaya Square. The Lukashenko government, which had vowed to crush any opposition in advance of the election, arrested hundreds of protesters and jailed the leading opposition candidate after the election. Lukashenko had learned the lesson from the Leipzig protests. The problem for the opposition was how to decide to protest in an environment where the state exerted that much control.

In May someone posting under the name by_mob used LiveJournal, a piece of blogging software, to propose a flash mob for the fifteenth of that month. The Minsk flash mob had little of Wasik's intentionally confounding feeling—the idea was simply that people would show up in Oktyabrskaya Square and eat ice cream. The results were one part ridiculous and three parts depressing; police were waiting in the square and hauled away several of the ice cream eaters, all while being documented in the now-standard pattern as other participants took digital pictures and uploaded them to Flickr, LiveJournal, and other online outlets. These pictures were in turn recirculated by bloggers like Andy Carvin and Ethan Zuckerman, political bloggers who cover the use of technology as a tool for social change. Images of a repressive Belarus thus spread far beyond the borders of Minsk. Nothing says "police state" like detaining kids for eating ice cream.

The ice cream mob was not an isolated incident. Flash mobs were held to protest the banning of the Belarusian Writers Union ("Show up at the Supreme Court, read books by the writers in the organization") and the closing of the newspaper *Nasha Niva* on the day it was to be shut down ("Gather in Oktyabrskaya, reading copies of *Nasha Niva*"). In the fall perhaps the simplest flash mob ever proposed took place: "Walk around Oktyabrskaya smiling at one another." This action produced the same reaction from the state; attendees reported that the police were using the presence of a pocketknife to try one of the smilers with weapons possession.

The police weren't reacting to the ice cream eating, reading, or smiling itself. The chosen behavior was intentionally innocuous, because the real message lay not in the behavior but in the collective action. After the postelection protests in March, any coordinated public gathering, especially in Oktyabrskaya Square, had a political dimension; mere evidence that Belarusian youth were operating in any organized way was both a threat and a rebuke to the state. The government has reason to worry: the historical lesson from Leipzig suggests that any forum for public expression is dangerous, because no matter how innocuous the original form of organization is, if the state is seen to tolerate it, it can become a forum for more focused discontent. The threat from a group eating ice cream isn't the ice cream but the group. The Lukashenko government is thus worried about coordinated ice cream eating—but if they have learned the lesson of Leipzig, why don't they just stop the mobs before they even gather? What good is having secret police if you can't spy on dissidents and disrupt their activities? With that strategy, after all, photographs of the police dragging people out of the main square are far less likely to show up all over the world.

Here is where the change in social tools since 1989 manifests itself. In Leipzig the early organization of the protests was fairly visible, and the protests themselves were fairly invisible. In June 1989, for example, the GDR canceled the entire Leipzig Street Music Festival, organized by independent citizen groups, and arrested all the participating musicians. The degree of advance planning required made the festival an easy target. Meanwhile the protests themselves were visible only to other Leipzigers, because the government had such tight control over media. The problem Lukashenko faces is that in the intervening years our social tools have made it possible for protesters to reverse the formula. Now the organization of group effort can be invisible, but the results can be immediately visible. Because the cost of sharing and coordinating has collapsed, new methods of organization are available to ordinary citizens, methods that allow events to be arranged without much advance planning. Because the mobs were proposed via weblog, the state had no way of keeping track of who had seen the plan. They could not break up the plot, since there was no plot; the event was proposed in public, so there was no secret information to uncover. Even if the government had the surveillance apparatus to know the identity of all the blog readers, it had no way of knowing which of them were planning to attend.

Using the state's reaction against itself is a kind of jujitsu. The protesters in Belarus believe that the government will be less willing to use force if it knows it is being observed by the outside world, particularly by Western Europe and the United States. As a result, the opposition wants to create widely observable protests, while the government wants to prevent such events from taking place or, failing that, to prevent documentation of those protests from being distributed widely. But with flash mobs the government can't intercept the group members in advance, because there is no group in advance: like the Mermaid Parade photographers, the group is latent until the event itself occurs, then is formed on the spot, as a result of the actions of the individual participants. (Also like the Mermaid Parade photographers, the by_mob proposer did not and could not know in advance who might show up.) By using public tools, the original proposers of the flash mob forced the state to react after the fact, but that's only half the battle. A protest isn't a protest unless it's public, and this is the second half of the change. Once the state does react, the flash mob attendees can document and publicize the proceedings, using cameraphones and photo-sharing websites that are much harder to control than traditional media. Even though there were only a few days between the announcement that *Nasha Niva* would be closed and its final day of publication, the opposition was able to get a few hundred people to turn out on that day. This speed of organization is accompanied by relative permanence of documentation. In late April 2006 someone going by the name freejul created a LiveJournal account. On the twenty-eighth, he or she posted pictures of the *Nasha Niva* flash mob, then another set of photos from a May 1 event in solidarity with political prisoners in Belarus. The last post from the account was on May 5, a little over a week after the first post, but the pictures are still there for all to see. Another advantage of blogs over traditional media outlets is that no one can found a newspaper on a moment's notice, run it for two issues, and then fold it, while incurring no cost but leaving a permanent record.

Because so many people have access to the Web, the Belarusian government can't stem the formation of flash mobs in advance, and because the attendees have cameras, it can't break up the mobs without inviting the very attention it wants to avoid. In this situation, the Belarusian government is limited to either gross overreactions (a curfew in Oktyabrskaya, a ban on ice cream or the internet) or to waiting for the mob to form, then disrupting it.

Such protests may not succeed in toppling the government. The Leipzig protests were driven by forty years of discontent, the Lukashenko government is not as all-controlling as the GDR was, and the West was considerably more committed to the fall of the USSR and its clients than it is to the democratization of Belarus. And all sorts of groups may use this technique. John Robb, author of *Brave New War*, calls the current generation of terrorists "Open Source Guerrillas" and notes all the ways they are adopting social tools and patterns to coordinate their efforts. Like the Belarusian protesters, the terrorist networks are less tightly integrated with one another and thus are harder to detect or intercept before they act. But whoever is using these tools, political action has changed when a group of previously uncoordinated actors can create a public protest that the government can neither interdict in advance nor suppress without triggering public documentation.

One might choose to bemoan the triviality of the culture of the developed world for using flash mobs for amusement and distraction (the love rug) rather than for political engagement. This judgment is accurate enough, but only because it is a restatement of the original observation, that people with more at stake are making more of these tools. Why? Social tools create what economists would call a positive supply-side shock to the amount of freedom in the world. The old dictum that freedom of the press exists only for those who own a press points to the significance of the change. To speak online is to publish, and to publish online is to connect with others. With the arrival of globally accessible publishing, freedom of speech is now freedom of the press, and freedom of the press is freedom of assembly. Naturally, the changes occasioned by new sources of freedom are most significant in less free environments. Whenever you improve a group's ability to communicate internally, you change the things it is capable of. What the group does with that power is a separate question.

## Replacing Planning with Coordination

Blitzkrieg, or "lightning war," is one of the few military strategies that nonhistorians know by name. The vision of the German Panzer tanks bearing down on hapless French defenses in May 1940 is etched in communal memory; from the initial German victory it took only six weeks for France to surrender. As pervasive as the image of German strength and French weakness is, however, much about it is misleading. In the 1930s the German army was smaller than France's (a condition forced on it at the end of World War I), and by 1940 Germany was bankrupt in all but name; its fearsome Panzer III and IVs, the key to blitzkrieg, were in many ways inferior to the French Char Bs that they would encounter. To allow the Germans to carry the day so decisively, something other than guns and armor was in play.

Although the Panzers had smaller guns and less armor, they came equipped with one thing the French tanks didn't have: radios. We don't often think of a radio as a weapon of war, but the radios allowed the Panzer commanders to share information and make decisions in the heat of battle, while the French, limited in their communications with their tank commanders, were hampered in their information gathering. This disadvantage sharply curtailed their ability to react to changes in the battlefield. The radios transformed the Panzers from stand-alone pieces of military hardware into a kind of coordinated group weapon.

One reason the French tanks, with their superior weapons and armor, didn't carry the day, not even with the natural advantage of a defensive position, is that they couldn't process information as fast as the Germans. Would the course of the twentieth century have been radically different if the French tanks had had radios? It's always dangerous to imagine alternative histories because of the number of variables involved, but radios were being installed in the French tanks in the spring of 1940, as the Germans struck. If the French had installed the radios a month earlier or if the Germans had attacked a month later, would the French have prevailed?

It's unlikely, because the Germans brought a second advantage to the battlefield; they understood what radios were good for. The French regarded the tank as a mobile platform for accompanying foot soldiers. The Germans, on the other hand, understood that the tank allowed for a new kind of fighting, a rapid style of attack that required a much higher degree of autonomy among the commanders and a much higher level of coordination in the field.

The ability to turn a collection of tanks into a coordinated force rested on two very different kinds of things, in other words. First, it required the media with which to coordinate the tanks. No radios, no blitzkrieg. Second, it required a strategy that took the new possibilities into account. No new strategy, no blitzkrieg either. Neither the technological change nor the strategy alone was sufficient to ensure German victory, but together they changed the way the world worked.

Contrary to its image as involving an overwhelming German force, blitzkrieg was in fact a strategy for using a smaller but more nimble force against a well-provisioned opponent. It used the same advantage that the Belarusian protesters are taking advantage of—a social tool that allowed them to act collectively. Though the flash mob is a relatively new addition to the repertoire, the ability of weak groups to coordinate their actions against strong ones is the hallmark of much political action. In 1999 the Falun Gong, a Chinese religious organization, astonished and terrified the Chinese government by assembling ten thousand people in Zhongnanhai, a secure complex in Beijing where many of China's leaders reside. The gathering was peaceful, but its execution stunned the Chinese government, as it had had no idea it was coming, having been organized by text messages via mobile phones. Howard Rheingold, in *Smart Mobs*, documented an event in the Philippines in which thousands of outraged citizens quickly coordinated a protest in Manila after President Joseph Estrada's government voted to weaken his corruption trial. The rapid assembly of thousands of Filipinos in the streets, who had forwarded text messages advising people where to go and exhorting them to "Wear Blck," convinced the government to let the trial go forward, thereby dooming Estrada. In Spain, after the ruling Partido Popular (PP) wrongly blamed Basque terrorists for the horrific bombing of the Madrid transit system, the opposition rallied to turn it out of office, forwarding the text-message-friendly phrase "Who did it" from phone to phone.

Most of us have seen this kind of shift away from advance planning with the adoption of mobile phones. As mobile phones have spread, people have shifted to making less definite plans. We no longer say, "I'll meet you at six at Thirty-third and Third," we say, "Give me a ring when you get off work," or, "I'll call when I get to the neighborhood." Text messaging allows whole groups to experience that shift as well. The previous political examples demonstrate the growing ease of that kind of coordination. Falun Gong, as a membership organization, still had some of the advantages of central coordination. The Filipinos lacked that degree of cohesion, but they had been witnessing the months-long spectacle of Estrada fighting corruption charges. In Spain only four days passed between the bombings and the election, which the PP had been widely tipped to win. The more ubiquitous and familiar a communications method is, the more real-time coordination can come to replace planning, and the less predictable group reactions become.

## Angry Passengers, Faster Action

On January 3, 1999, Northwest Airlines flight 1829 took off from Miami on its way to Detroit. Flight 1829 ordinarily flies from the Caribbean vacation spot of St. Martin, but the day before, because of a snowstorm in Detroit, it had been diverted to Miami. The flight left Miami a little after noon and landed in Detroit at 2:45 p.m. The passengers' trip that day was less than a third over.

Though the snow had stopped falling, the Detroit airport had been unprepared for the storm. The additional flights from the previous day's closure, the snow still to be cleared, and insufficient staff all meant that not enough gates were open. After Flight 1829 landed, the pilot was directed to pull over to a side runway, and the passengers were told to expect a two-hour delay, which had the predictable effect on their mood. Two hours came and went with no gate clearance; the flight attendants struggled to keep the passengers mollified with limited supplies. They had not stocked up on food or drinks in Miami (it was to have been a short flight) and were running out of liquor as passengers continued drinking to dull their annoyance. Three hours passed, then four. The lavatories began to smell, then clog, then leak. Lawyers on board were signing up potential plaintiffs. Passengers with babies, heart conditions, and nicotine habits all pleaded with the flight crew to get them off the plane. These pleas were forwarded to the flight deck, who in turn called the ground crew, who offered little more than assurances that they knew things were bad and were working on it.

Five hours passed. Flight attendants began encouraging passengers to write letters of complaint to the CEO. Someone suggested calling him instead. They found his name, John Dasburg, in the in-flight magazine, and his home phone number via directory assistance. They called his house. He wasn't home, but his wife answered and got an earful from the passengers. The captain, learning that a passenger had called Dasburg, summoned the caller to the cockpit and asked for the number. The captain himself then called Dasburg to demand that a gate be opened. That—finally—got results. The plane pulled out of the line (to the understandable frustration of the other waiting pilots) and headed to the newly opened gate. At 9:42 p.m., the passengers finally disembarked, seven hours after they'd landed.

This tale resulted in incredibly bad press for Northwest and for the airline industry generally. The net result, though, was negligible. If any letters of complaint were delivered to

Dasburg, they produced no apparent change. The lawsuit, for "false imprisonment and breach of contract," was settled out of court, and the airlines adopted a toothless and voluntary Customer Service Initiative (which should have been redundant, given the business they are in). People had been subjected to quite incredible torment from a company nominally in the business of providing a service, but in the end the power in that particular situation lay with the airline, not with its customers.

Precisely this imbalance of power made what happened next time so remarkable.

The flight numbers, cities, and dates were different, but the basic story was the same. On December 29, 2006, several American Airlines flights were diverted to Austin because of heavy storms in Dallas. Once the planes were on the ground, they waited for hours, with cockpits powerless to obtain gates, increasingly agitated passengers, insufficient food and water, and overflowing toilets. It was a replay of Detroit, minus the wind chill but with additional delays—after landing some flights sat on the ground for more than eight hours before the passengers were let off.

Kate Hanni, a real estate agent from California and a passenger on American flight 1348, got angry. In the days after the delay she formed a group to represent the rights of passengers. They proposed an Airline Passengers' Bill of Rights (sample item: "Provide for the essential needs of passengers during air- or ground-based delays of longer than 3 hours"), they lobbied Congress (adoption of the passengers' bill was proposed in the House and Senate as a result), and they invited the general public to sign their petition; they garnered thousands of signatures within weeks. Partly as a result, every new airline horror story—like the epic tarmac delays suffered by JetBlue passengers on Valentine's Day 2007, or the eight-hour delay of another American Airlines flight in April—are now covered by the press as part of a larger issue, rather than just a single event, spreading awareness even further. After the JetBlue meltdown, the CEO stepped down, and the company adopted its own Passenger Bill of Rights.

The results of the ground delays in Detroit and Austin could hardly have been more different. In Detroit the cumulative fury of the passengers, despite their mistreatment, dissipated quickly. In Austin, the fury drove the creation of an organization within days that quickly went national and had an almost immediate impact, changing the legislative agenda, press coverage, and public expectations of the airline industry. The Detroit passengers were as badly treated, and as angry about it, as the Dallas passengers. Why did one infuriating delay lead nowhere, while the other led to a real increase in pressure on the airlines?

The key change was that Kate Hanni had in her hands the tools to encourage and sustain participation. She had the desire to do something, and in 2007 she was able to communicate that desire in a way that created a public movement, using tools that have become commonplace.

It started with a simple conversation. While she was searching the Web for details about the flight, Hanni found a short story about the delays in an Austin newspaper, the *American-Statesman*. She posted two separate comments on the article, spelling out in great detail what had happened to the passengers that day. (The combined length of these comments was over four times the length of the original story.) At the end of her second comment she wrote, "Anyone from this flight please contact me."

Another passenger on the flight responded directly to Hanni and offered contact details for additional passengers. Once the *American-Statesman* allowed comments by readers, what had previously been a one-way platform (journalist talks to readers) became first a shared platform (Hanni offers her observations to the world) and then a cooperative platform (Hanni uses the article as a means to communicate with other passengers). Within days she had contacted enough people to put together a group with a mission and a name—Coalition for an Airline Passengers' Bill of Rights. She created an online petition; more than two thousand people signed it in the first month, a number far in excess of the number of passengers who had been directly affected. Hanni and other coalition members were interviewed by the *New York Times*, CNN, and CBS, and a variety of travel websites linked to the coalition weblog. All of this attention created not just awareness but the possibility for new action—new signatures, new calls to Congress, new donations.

Just as social tools are creating members of the former audience, they are creating legions of former consumers, if by "consumer" we mean an atomized and voiceless purchaser of goods and services. Consumers now talk back to businesses and speak out to the general public, and they can do so en masse and in coordinated ways. The U.K. division of the bank

HSBC had been recruiting students and recent graduates with the promise of checking accounts that carried no penalty for overdrafts. In August 2007, HSBC decided to revoke this policy, giving the students only a few weeks' notice of the impending change. The move obviously made corporate sense; interest-free borrowing was costing HSBC money, and the so-called switching costs for the students—the costs of finding another bank and transferring accounts—would decrease the likelihood of mass defections. Having used the interest-free overdrafts to attract customers, HSBC reasoned that it could cancel the program with little penalty.

HSBC hadn't reckoned on Facebook, the social networking service that started its life specifically targeted to college students. A Cambridge University student and vice president of the student union named Wes Streeting set up a place on Facebook to complain about the policy, calling it "Stop the Great HSBC Graduate Rip-Off!" Similar to the Flyers Rights story, thousands of students signed up in a matter of days. Critically, Facebook was the one place where both current students and recent graduates could all be reached together; in years past, the dispersal of the graduates made it hard to communicate with them, but now they remain part of the social fabric of a college even after dispersing physically. Facebook also helped lower the switching costs, as current and former students began researching and recommending other U.K. banks that still offered interest-free overdrafts. As a result of better information about the alternatives, and because individual actions could be part of a larger movement, far more students began publicly threatening to move accounts than HSBC had anticipated.

Seeing the large and growing response, the Facebook group then announced it would stage a public protest at HSBC's offices in London in early September. That protest never happened, for the simple reason that HSBC caved in long before the appointed date; seeing the online protest and threatened with a real-world one, and having underestimated what pooled information would do to switching costs, HSBC reversed the policy. Andy Ripley, the head of product development, explained the reversal by saying, "Like any service business, we are not too big to listen to the needs of our customers." Though face-saving, this is a curious statement, since HSBC could have predicted the students' unhappiness with the change long before implementing it. The reversal didn't come about because the students were unhappy; it came about because they were unhappy and coordinated.

The enormous effect of motivation is obvious—the Flyers Rights and the HSBC protests relied on Hanni and Streeting to get them started. Less obvious but equally important is the limited motivation of most of the participants in the protests. Many people care a little about the treatment they get from airlines or banks, but not many care enough to do anything about it on their own, both because that kind of effort is hard and because individual actions have so little effect on big corporations. The old model for coordinating group action required convincing people who care a little to care more, so that they would be roused to act. What Hanni and Streeting did instead was to lower the hurdles to doing something in the first place, so that people who cared a little could participate a little, while being effective in aggregate. Having a handful of highly motivated people and a mass of barely motivated ones used to be a recipe for frustration. The people who were on fire wondered why the general population didn't care more, and the general population wondered why those obsessed people didn't just shut up. Now the highly motivated people can create a context more easily in which the barely motivated people can be effective without having to become activists themselves.

## Banal Tools in Remarkable Contexts

Evan Williams is a natural inventor of social tools. In the 1990s his company, Pyra, was working on a complex project-management tool that they could sell to businesses, but while doing so, they needed a project-management tool for themselves. Instead of simply adopting their own tool (which wasn't ready and was in any case too complex for a little company), they wrote just about the simplest application one could imagine. It was a website that would take text that a user entered into a form, and post it onto a webpage, with the most recent additions at the top of the page. The tool, simple as it was, turned out to be far more compelling than the software they were supposed to be creating, and they ended up working more on the in-house tool than on their nominal product. They named their product Blogger and launched it to the world. It spread like wildfire—hundreds of thousands of users adopted it within a few months. (Blogger was ultimately acquired by Google.)

Evan's next idea was audio blogging, where users would

post short recorded bits of sound to a website, to be listened to by others. This idea didn't take off in the same way, but it did get him focused on mobile phones. His next idea relied on text messaging, the short written messages many people can send from their mobile phones. His service, called Twitter, was simplicity itself. To use Twitter, you create an account for yourself, and then you send Twitter a message, via the Web, by instant message, or from your phone. A message on Twitter, called a tweet, is a short snippet of text, usually an update about what you are doing; sending a tweet is "twittering." The message goes to your friends who are also on Twitter and, if you like, gets posted to the Twitter "public timeline," a webpage with the most recent public twitters.

Much of the content on the public timeline is inane. On a random Saturday afternoon, here's a random sample of twittering:

> jmckible says "Just had to blow out my DS slot NES style"
>
> truejerseygirl says "Hosting a CD Exchange party tonight. Made the jello shots, bought the booze and chips, but havent burned all the cd's yet. Eek Im a slacker"
>
> laurence says "At Maker Faire"
>
> Josh Lawrence    In friggin' heaven because I'm eating Trader Joe's gourmet chocolate fudge.
>
> Mike Barrett    mrmanager07 WOOO summer courses are FUUUUN

Many of the public posts have this sort of quality—video games, pop music, and Jell-O shots—where the publicly available content is not likely to interest most users. Like weblogs that are written for small clusters of friends, most twittering is for the benefit of friends rather than for the general public. These twitters are interesting not so much because the messages themselves are informative, but because the receiver cares about the sender. You probably don't care that laurence is at the Maker Faire (a Silicon Valley event for the DIY movement), but if you knew laurence, or were at the Maker Faire yourself, you might. As always, socially embedded messages are more valuable than random public broadcasts. Even accepting that Twitter creates a kind of peripheral vision for what someone's friends are doing, though, it can seem awfully banal. Until you see Alaa's feed.

> Going to doky prosecutor judge murad accused me and manal of libel 10:11 AM April 04
>
> Waiting for prosecutors decision might actually spend the night in custody 01:57 PM April 04
>
> We are going to dokky police station 03:31 PM April 04
>
> In police station no senior officers present so we are in limbo 04:29 PM April 04
>
> We will not be released from giza security will have to go back to dokki station 07:59 PM April 04
>
> On our way back to police station 10:25 PM April 04
>
> We are free 11:22 PM April 04

Alaa Abd El Fattah is an Egyptian programmer, democracy activist, and blogger living in Cairo. Here he is documenting his arrest, with his wife Manal, in El Dokky, a Cairo neighbor-hood, an episode that ended twelve hours later with their re-lease. His arrest was ordered by Abdel Fatah Murad, an Egyptian judge who was attempting to have dozens of websites blocked in Egypt, on the grounds that the sites "insult the Quran, God, The President and the country." When Egyptian prodemocracy bloggers started covering the proposed censorship, Murad added their sites to the list he was attempting to ban.

What does a service like Twitter, whose public face is so banal, offer Abd El Fattah and the other Egyptian activists? Some of the value is fairly prosaic—free speech activists are harassed or detained in several countries in the Middle East, so they use Twitter to alert one another as to whether they've passed through various security checkpoints (often at air-ports); the absence of a message may mean they've been de-tained. On other occasions, though, it provides a way to spread real news. Here is how Alaa reported the news of the arrest and continuing detention of Abdel Monem Mahmoud, an-other Cairo blogger.

> they've arrested ikhwani blogger monem (http://
> ana-ikhwan.blogspot.com) we must organize a cam-
> paign 10:07 AM April 13
> turns out Monem did not turn himself in yet, he is
> hiding from police until lawyers find out more de-
> tails, but they did break into his home 03:31 PM
> April 13
> In case you dont know momen got arrested early
> today at cairo airport 04:17 PM April 15
> monem appeared before shobra prosecutor and he

will be detained for 15 days. 07:50 PM April 15

Monem and co start hunger strike due to maltreat-
ment 03:36 PM May 07

Abd El Fattah and Mahmoud do not see eye to eye
politically—Abd El Fattah is a secular blogger, while Mahmoud
is a member of the conservative Muslim Brotherhood—but
both have an interest in free speech, and these tools allow
citizens to report the news when they see it, without having to
go through (or face delay and censorship by) official news
channels. Twitter also offers an ability to coordinate these
users' reactions to the state. As El Fattah describes Twitter,
"We use it to keep a tight network of activists informed about
security action in protests. The activists would then use twitter
to coordinate a reaction." Because prodemocracy activists are
watched so carefully, Twitter allows them a combination of
real-time and group coordination that helps tip the balance of
action in their favor. One early use of Twitter had El Fattah and
a dozen or so of his colleagues coordinating movements to
surround a car in which their friend Malek was being held by
the police, to prevent it and him from being towed away.
Knowing they were being monitored, they then sent messages
suggesting that many more of them were coming. The police
sent reinforcements, surrounding and thus immobilizing the
car themselves. This kept Malek in place until the press and
members of Parliament arrived. The threat of bad publicity led
to Malek's release, an outcome that would have been hard to
coordinate without Twitter.

The power to coordinate otherwise dispersed groups will
continue to improve; new social tools are still being invented,

and however minor they may seem, any tool that improves shared awareness or group coordination can be pressed into service for political means, because the freedom to act in a group is inherently political. The progression from Leipzig to increasingly social and real-time uses of text messaging from Beijing to Cairo shows us that we adopt those tools that amplify our capabilities, and we modify our tools to improve that amplification.

# CHAPTER 8

# SOLVING SOCIAL DILEMMAS

*There are real and permanent social dilemmas, which can only be optimized for, never completely solved. The human social repertoire includes many such optimizations, which social tools can amplify.*

Let's say, for the sake of illustration, that you and I went out for a few drinks last Saturday night, and at around 2 a.m. one of us said, "Hey, I know! Let's steal a car!" (I think it was you who said that.) So we steal a car, one thing leads to another, mistakes are made, and half an hour later we crash right through the window of a store. We barely have time to jump out and pretend to be bystanders before the police arrive.

Now the police aren't really buying the bystander alibi, but they don't have any other witnesses, so they take us off into separate rooms for questioning. Once we are separated, they make each of us this offer: "Look, we think you're innocent, but we suspect the other person in the car was responsible. If you tell us what you know about them, we'll give you a big reward, and file charges against them. But you gotta tell us right now, and if you don't, we're going to hold you overnight." Since each of us is getting this offer, it creates four possibilities:

1. We each stick to our stories, they've got no evidence, and they keep us both overnight.
2. I stick to the bystander story and you turn me in. You get a reward, while I get charged.
3. I turn you in while you stick to the story. I get a reward, while you get charged.
4. We turn each the other in. We both get charged.

So knowing that I face the same choice as you—sticking to my story or turning you in—what do you do?

The worst outcome would clearly be getting charged with a crime, and the best outcome would be getting the reward. You know that I know that too, and if we both try to get the reward, we both get charged. The second best outcome is spending the night in jail, but you know that I know that too, and if you stick to your story in an attempt to get this outcome, I can go for the reward by turning you in. Similarly, if I stick to my story in an attempt to get the night in jail, you can turn me in to try to get the reward, but if we both try to get the reward, we both get charged—back to the worst outcome again.

This is a simplified version of the Prisoners' Dilemma, a social science thought experiment about how people make decisions. (The payoff matrix is bit more complex in the standard version, but the dilemma is the same.) Assuming that the two people can't communicate with each other and don't trust each other (about which more in a moment), the worst outcome—number four—is the rational one, an outcome called a Nash equilibrium. The dilemma of the Prisoners' Dilemma is that, because it is a one-off transaction in which you and I can't communicate with each other, we can't coordinate any outcome better than the dismal Nash equilibrium. (This is the same math underlying the Tragedy of the Commons, where the Nash equilibrium encourages individual defection, even as it damages the group.) Things change, though, when the prisoners interact with each other repeatedly, a version called an iterated Prisoners' Dilemma.

Robert Axelrod, a sociologist at the University of Michigan who studied the iterated version extensively, staged tournaments for different software programs emulating the prisoners. Each program was given a strategy for when to cooperate and when to defect (the same two choices you and I faced in our notional interrogation rooms). These strategies were measured by adding or deducting points for the various outcomes. After running the tournament with many different participating strategies, ranging from "always defect" to "cooperate or defect at random," Axelrod found that a single strategy, called Tit-for-Tat, was most successful against every other strategy tried. Tit-for-Tat started by trying to cooperate the first time it was paired with any other program. If that program also cooperated, then Tit-for-Tat would offer to cooperate in the next round, and so on. As long as another program offered to cooperate, Tit-for-Tat would continue to do so as well. If the other program defected, though, taking advantage of Tit-for-Tat's trusting behavior, then Tit-for-Tat would defect against that program in the next round, effectively punishing the other program as a way of communicating that its trusting nature extended only to those who reciprocate.

This strategy is a highly simplified version of real life—the more general lesson is that people who interact with one another repeatedly communicate through their actions, introducing what Axlerod calls "the shadow of the future." We all face the Prisoners' Dilemma whenever we interact with people we could take advantage of, or people who could take advantage of us, yet we actually manage to trust one another often enough to accomplish things in groups. The shadow of the future makes it possible for me to act on your behalf today, even at some risk or cost to me, on the expectation that you will remember and reciprocate tomorrow.

## New Tools to Create Social Capital

Over on University Place in lower Manhattan, a few blocks from my office, is the local bowling alley. Bowling often conjures up an era of picket fences and twenty-five-cent Cokes, and our local bowling emporium even has a name reminiscent of that time—Bowlmor Lanes. On any given Friday night, though, Bowlmor is very much an institution of the moment, catering to martini-sipping twentysomethings instead of factory workers unwinding with a beer. Through the decades bowling has been persistently reinvented, and it remains a durably popular activity. But between the 1950s and now there has been one significant change—a precipitous decline in league bowling, with its memberships and seasons and uniforms and all the rest. Though plenty of groups bowl at Bowlmor Lanes, they are mainly people who already know one another; the bowling is more a consequence of group interaction than a source of it. The gradual disappearance of bowling leagues is one of many reductions in social mechanisms whereby people may be introduced to one another as a consequence of shared activity. This doesn't matter much for the fate of Bowlmor Lanes—a customer is a customer, league or no—but it may matter for the country.

When Robert Putnam, a Harvard sociologist, published *Bowling Alone* in 2000, it was an immediate sensation. His account of the weakening of community in the United States, based on a huge number of indicators from the decline of picnicking to the abandonment of league bowling, offered two provocative observations. First, much of the success of the United States as a nation has had to do with its ability to generate social capital, that mysterious but critical set of characteristics of functioning communities. When your neighbor walks your dog while you are ill, or the guy behind the counter trusts you to pay him next time, social capital is at work. It is the shadow of the future on a societal scale. Individuals in groups with more social capital (which is to say, more habits of cooperation) are better off on a large number of metrics, from health and happiness to earning potential, than those in groups with less social capital. Societies characterized by a high store of social capital overall do better than societies with low social capital on a similarly wide range of measurements, from crime rate to the costs of doing business to economic growth.

This is the shadow of the future at work: direct reciprocity assumes that if you do someone a favor today, that person will do you a favor tomorrow. Indirect reciprocity is even more re-markable—it assumes that if you do someone in your commu-nity a favor today, someone in your community will be around to do you a favor tomorrow, even if it isn't the same person. The norms and behaviors that instantiate the shadow of the future is social capital, a set of norms that facilitate cooperation within or among groups.

It was Putnam's second observation, however, that gener-ated the real reaction. Across a remarkably broad range of measures, participation in group activities, the vehicle for cre-ating and sustaining social capital, was on the decline in the United States. Putting the two observations together, he con-cluded that one of the greatest assets in the growth and stabil-ity of the United States was ebbing away. One cause of the decline in social capital was a simple increase in the difficulty of people getting toge̶ ̶ ̶ ̶ ̶ ̶ ̶ ̶ ̶ ̶ ̶ ̶transaction costs, to use Coase's term.

sive, either in direct

of it, and several e̶

smaller households

the spread of televis̶

the transaction cos̶

work. For most pe̶

conclusion was n̶

ice cream socials.

nity. In the 1990

successful web b̶

for his next busi̶

of regarding it a̶

trying to reinvigorate the creatio̶ ̶ ̶ ̶

*Handwritten note:*

COLLECTIVE ACTION.

Ex): 143–160

SUMMARY: 47–54

Ratio & Simple Group Formation: 151–153

Sharing Info: 148–151

Ex) 51–53, 135, 137, 190, 275

As a Type of Group Action: 51–53

real-world interaction. The solution he came up with was surprisingly simple.

First Heiferman assumed that people knew what they were missing and would want it back if they could get it; in an era of declining social capital, people would take steps to increase their communal participation if someone could make it easy again. Second, he recognized that treating the internet as some sort of separate space—cyberspace, as it was often called—was part of the problem. That word, coined by William Gibson in his novel *Neuromancer*, refers to a kind of alternate reality mediated by the world's communications networks. The cyberspace of *Neuromancer* is a visual representation of all the world's data; John Perry Barlow, a digital rights activist, later used the word to refer to the social spaces of the internet. Whether visual or social, though, the basic sense of cyberspace was that it was a world separate and apart from the real world. The predicted end point of this process was a progressive disassociation of social life from real space, leading to the death of cities as the population spread out to more bucolic spots.

The assumption that communications tools are (or will someday be) a good substitute for travel assumes that people mainly gather together for utilitarian reasons of sharing information. Companies have been selling us this idea since the invention of the telegraph, and AT&T's famous Picturephone, first launched at the 1964 World's Fair, was pitched as a way to reduce the need for travel. This reduction did not happen, not in 1964 or ever. If communication were a substitute for travel, then the effects would have shown up by now, but they haven't. In 1978 President Carter deregulated the airlines, causing travel prices to fall, but telecommunications stocks didn't collapse; they rose. Similarly, in 1984 Judge Harold Greene broke up AT&T, leading to a rapid decrease in long-distance phone call costs; airline customers increased that year. Communication and travel are complements, not substitutes. Chris Meyer, a globe-trotting consultant for the Monitor Group, observes that "better communications make it easier for me to keep in touch with the office, so I spend *more* time on the road, talking to clients."

We gather together because it is useful but also because we like to. Assuming that videophones or e-mail or virtual reality will reduce the overall amount of travel is like assuming that liquor stores will kill bars, since liquor stores sell drinks much more cheaply than bars do. In fact, the reason people go to bars is not simply to get a drink, but to do so in a convivial environment. Similarly, cities don't exist just because people have had to be nearby to communicate; cities exist because people like to be near other people, and it is this fact, rather than the mere trading of information, that creates social capital. (Anyone who predicts the death of cities has already met their spouse.) This obvious human preference was overlooked during the early public spread of the internet, in large part because the average user interacted with different people online and offline.

What seemed like a deep social change in the 1990s was revealed to be a temporary accident by the year of Meetup's founding. The idea of cyberspace made sense when the population of the internet had a few million users; in that world social relations online really were separate from offline ones, because the people you would meet online were different from the people you would meet offline, and these worlds would rarely overlap. But that separation was an accident of partial adoption. Though the internet began to function in its earliest form in 1969, it was not until 1999 that any country had a majority of its citizens online. (Holland was first, but that condition now applies to most countries in the developed world.) In the developed world, the experience of the average twenty-five-year-old is one of substantial overlap between online and offline friends and colleagues. The overlap is so great, in fact, that both the word and the concept of "cyberspace" have fallen into disuse. The internet augments real-world social life rather than providing an alternative to it. Instead of becoming a separate cyberspace, our electronic networks are becoming deeply embedded in real life.

Heiferman realized that if enough people are online, you don't have to group them solely by affinity (pug lovers, White Stripes fans, libertarians, whatever). Instead you can group them by affinity *and* proximity (pug lovers in Poughkeepsie, White Stripes fans in Walla Walla). He designed Meetup to help people find each other online and then meet in the real world, taking the burden of coordination off the hands of the potential users. Meetup users can search by interest (Are there any relevant Meetups in my town?) or they can look by area (I live in Milwaukee, what Meetups are nearby?)

By registering people's interests and location, Meetup can identify latent groups and help them come together. Heiferman bet that all over the United States (and later, the world) latent groups would be happy to get together if someone solved the coordination problem. Armed with this intuition (and the work of a talented group of programmers and designers), he launched the service. In early talks to potential users or inves-

tors he sometimes presented Meetup as a kind of time machine, reinvigorating classic American interest groups—people
who shared an interest in bowling, cars, or Chihuahuas. (He
talked about people who liked Chihuahuas so often, in fact,
that it became a trademark bit of his spiel.)

The groups that actually ended up using Meetup didn't
look anything like Heiferman expected. Here's the list of the
fifteen most active Meetups the year after the site launched:

| Topic | Total Meetups | Total Members |
|---|---|---|
| Witches | 442 | 6,757 |
| Slashdot | 401 | 11,809 |
| LiveJournal | 311 | 10,691 |
| Bloggers | 136 | 4,222 |
| Pagans | 90 | 2,841 |
| Fark | 81 | 4,621 |
| Ex–Jehovah's Witness | 67 | 1,609 |
| Bookcrossing | 56 | 4,414 |
| Xena | 51 | 1,641 |
| Tori Amos | 47 | 2,261 |
| Ultima | 38 | 2,467 |
| Star Trek | 35 | 1,196 |
| Radiohead | 32 | 1,986 |
| Vampires | 28 | 1,339 |
| Atheists | 27 | 1,338 |

This list is unlike any list of American groups ever assembled. It measures something important (or rather it collates
several different important things) because it demonstrates

that Meetup's convening power lies nor in recreating older civic groups but in creating new ones.

The groups represented here can be divided into three broad categories. The first, including Witches, Pagans, Ex–Jehovah's Witnesses, and Atheists, are people who share some religious or philosophical outlook but have no support from the broader U.S. culture. There are many more Presbyterians than pagans in the United States, but the Presbyterians aren't on this list because they don't need Meetup to figure out when and how to assemble; they meet every Sunday morning at the Presbyterian church. Because they are both internally organized and externally supported, Presbyterians suffer less than pagans from transaction costs, who have no culturally normal place and time to meet and no ready way to broadcast their interests without censure. Jehovah's Witnesses enjoy advantages similar to those of other Christian sects, but ex-Witnesses turn to Meetup because they don't enjoy that kind of socially supported coordination.

The second category of Meetup groups includes the members of websites and services who would like to assemble with other users of those services in real life. This group includes Slashdot, LiveJournal, Bloggers, Fark, Ultima, and Bookcrossing. (Interestingly, the numbers show how clustered these groups are; though Slashdot and LiveJournal had more members than Witches did, they met in fewer cities; or put another way, Witches are more evenly distributed in U.S. society than are geeks or bloggers.) This is what the end of cyberspace looks like: the popularity of these Meetup groups suggests that meeting online isn't enough and that after communicating with one another using these various services, the members become convinced that they share enough to want to get together in the real world. Especially relevant to this thesis is the Ultima group. Ultima is an online game set in an imaginary world, Britannia, rendered in 3D, where players interact with one another. It is one of a class of games called "massively multiplayer online role-playing games," or MMOs for short. If virtual interactions were ever enough to be completely satisfying, we'd expect them to work best in these virtual worlds. But the popularity of Meetup groups for virtual contacts shows that even online communication that emulates face-to-face interaction still leaves people wanting real human contact.

The third category includes fans of cultural icons quirky enough that those fans want to be in one another's presence. LiveJournal users can at least potentially come in contact with one another on the website, but Tori Amos fans are simply guessing that they will get along. (The Vampires group falls into both the first and third categories.) To want to be in other people's company without having spoken before, on the basis of a shared cultural affinity, is a pretty good advertisement for Heiferman's initial thesis—that even in a mediated age, people crave real human contact.

These three categories have several things in common. First, they represent not just things people do but ways they think of themselves (and of other people). Many more people use Google than LiveJournal, but there is no broad interest in a Google users' Meetup group. Second, this self-conception translates into a desire to meet with other people who share the same interests. Many more people were watching *Everybody Loves Raymond* in 2002 than were watching *Xena: Warrior Princess,* but *Xena*-fandom was a better predictor of real commonality. Finally, the world provided no easy way for these people to find one another prior to Meetup. Because the audience for *Xena* was passionate but small, the likelihood that *Xena* fans would find one another at random was similarly small, but precisely because of this minority status, the likelihood that, once they did, they would feel some sense of kinship was higher than average. This effect is general. Lada Adamic, a researcher at HP Labs, studied the users of an online student center at Stanford called Club Nexus and found that two students were likely to be friends if their interests overlapped, and that the likelihood rose if the shared interests were more specific. (Two people who like fencing are likelier to be friends than two people who like football.) The net effect is that it's easier to like people who are odd in the same ways you are odd, but it's harder to find them. Meetup, by solving the finding problem, created an outlets for many new groups—groups that had never been able to gather before.

Meetup didn't end up recreating the old model of community, because it provided a different set of capabilities; the groups that took first and best advantage of those capabilities were the groups with a latent desire to meet but had faced previously insuperable hurdles. These groups aren't the classic American interest groups of yore; many of the most popular groups tell us surprising things about what our society is like right now.

## Stay at Home Moms and the Politics of Exclusion

One of the most popular current groups on Meetup is Stay at Home Moms (SAHM). Mothers with young children have been gathering in groups since before the invention of the internet, in fact before the invention of agriculture. This is an old pattern, so why would SAHM Meetups be so popular? The answer, in one sentence, is that modern life has raised transaction costs so high that even ancient habits of congregation have been defeated. As a result, things that used to happen as a side effect of regular life now require some overt coordination.

Some of the hurdles to be overcome are physical. As of the 2000 census, a majority of the U.S. population lived in the suburbs, and in the suburbanized United States, physical distance raises several barriers. Houses are often separated from commerce, so much of the time spent doing errands or ferrying children from hither to yon is spent in a car. In a pedestrian setting, running into someone is a good thing; in a car, not so much. Both the distance between the grocery store and home, and the fact that travel between the two is highly enclosed, reduce the likelihood of chance social encounters (and as a result reduces the raw material for building social capital).

As the two-income family has become more normal, the center of gravity for social interaction has shifted from the neighborhood to the workplace. Not only have the suburbs reduced the likelihood of chance encounters, but the increased percentage of the population with jobs, including especially a sharp increase in the number of women, means that the workplace now has many of the characteristics that the neighborhood used to have. You are likelier to be introduced to new coworkers than to new neighbors, and interactions at work produce the kind of familiarity and trust that used to be more a part of the fabric of our communities.

Meetup makes the coordination of groups simple, offering

a way of undoing at least some of the damage inflicted on that fabric. This is one reason groups like Stay at Home Moms matter so much. Some groups we expect to be technology-obsessed; maleness, singleness, and youth all correlate with technophilia, while femaleness, age, and family life don't. So when a group of mothers adopts a piece of technology, it indicates an expression of preference far more serious than seeing a thirteen-year-old boy go wild over an Xbox. The popularity of groups like Stay at Home Moms indicates that Meetup's utility in helping people gather in the real world is valuable enough to get the attention of people who are too busy for most new tools.

The most successful Meetup parents' group didn't turn out to be the most general one. Meetup also lists a Parents and Kids Playgroup, which describes a much larger class of potential members than Stay at Home Moms does, but the Parents and Kids group is significantly less popular. This is one of the essential conundrums of social capital—inclusion implies exclusion. The very name Stay at Home Moms is a salvo in the decades-long conversation about the ideal structure of a family—this group is for mothers who are playing a relatively traditional role in child-raising. Though it is hard to imagine a man with a child being turned away from the North Charlotte Stay at Home Moms Meetup, say, it's also hard to imagine that a lot of dads show up in the first place.


## Self-Help We Don't Approve Of

In 2002 I taught a graduate course at New York University called "Social Weather," about the experience of participating in online groups. The course's title was an analogy to the way the weather affects our mood; in the class we were looking at how social groups create an emotional environment that affects all the participants. One of my students in that class, Erika Jaeggli, was also working on the magazine *YM*'s website. *YM* (formerly *Young Miss*, then *Your Magazine*, then just *YM*) is designed to appeal to teen girls. In 2002, like almost every other magazine in the country, *YM* was wrestling with how to embrace the Web. In addition to putting the magazine's articles online, the staff created a set of online bulletin boards where *YM* readers could go online and talk to one another about whatever was on their mind. Popular topics included clothes, school, romance, and health and beauty—pretty standard fare for teen girls. Erika's job was half host, half chaperone, working to draw the girls out and make them feel comfortable talking to one another, while also keeping the conversation from devolving into name-calling or turning to inappropriate subjects. Particularly at an age when readers were exploring previously off-limits subjects like sex or the use of alcohol and other drugs, the role of an editor was a balancing act. Too little intervention, and the conversation would turn into bedlam; too much would seem like a ham-handed attempt to bring the girls into line—precisely the kind of treatment from adults they were coming to the *YM* website to escape.

A few months into the semester Erika stopped me in the hallway to tell me *YM* was shutting down its health and beauty bulletin board. When I expressed surprise that a magazine focused on teen girls would kill off those discussions, she said, "Most of the girls were fine, but we couldn't figure out how to stop this one group of girls from swapping tips on remaining .

anorexic." These Pro-Ana girls (short for pro-anorexia) were posting pictures of models and actresses whose rib cages were showing as "thinspiration" and exhorting each other with "You've made a decision—you won't stop. The pain is necessary, especially the pain of hunger. It reassures you that you are strong—can withstand anything—and that you are NOT a slave to your body; you don't give into its whining."

Most dangerously, the Pro-Ana girls were trading practical advice (though the word "practical" is odd in this context):

> You can train yourself to forget hunger by gently punching your stomach every time you get hungry because you'll hurt too bad to eat.

> Take TUMS to help with hunger pains; they have calcium so they'll help in that area also.

> Clean something you find truly disgusting. Afterwards, you won't feel like eating for another couple of hours.

The problem for *YM* wasn't that the bulletin board had failed to get the interest of their readers. The problem was that it had succeeded in a way for which *YM* was unprepared.

Whenever individuals want to find one another, the larger society in which they are embedded can provide or withdraw support for their association. Much of the way we talk about identity assumes it is a personal attribute, but society maintains control over the use of identity as an associational tool. A recovering addict would find it very risky to ask coworkers for help finding a support group, as might someone looking for the local gay community. Whether society offers or withholds this support, however, matters less with each passing year.

Here is the dilemma the *YM* staff found themselves in. To host a conversation among their most active and engaged readers, they had to monitor the site, but if Erika and the other online editors had weeded out every mention of anorexia, they would come to seem like bullies, especially as some of the conversations were genuinely about avoiding anorexia. Further complicating things, the Pro-Ana girls were willing to go to great lengths to have their discussions out in the open. In the end, the possible sweet spot between too little intervention and too much came to seem illusory, and *YM* simply shut down the conversation, rather than engage in daily censorship or risk having the girls who congregated at *YM* get sick. But what exactly had the girls done that presented such a novel challenge? Anorexia has been a source of public worry since the 1960s, and groups of girls have been hanging out together for decades, talking about everything from sex and drugs to fashion and food. Did *YM* just act on the standard fear that new technology would bring ruin to society? Or is something different?

Something is different. It is easier for groups to form without social approval. Predictably, the Pro-Ana movement has simply moved from hosted conversation spaces like that on *YM* to more open tools like weblogs and social networking sites like MySpace. *YM* was able to withdraw its support for the group on its own site, but neither it nor any other organization could prevent the girls from forming groups and conversing with one another if they wanted to. Before we had any real group-forming technologies, merely finding people who were interested in the same things was hard, and most of the ways we had for doing so—from putting up flyers around the neighborhood to taking out an ad in the local paper—were expensive and time-consuming. Because of these difficulties, social approval could make group-forming much easier, and social disapproval could make it much harder. Formal mechanisms like the law are one factor: it is easier to find a group of people to drink with than to shoot up with, because the law treats alcohol and heroin differently. But legal strictures account for only a small number of these cases; there are many more informal mechanisms for creating the same effect.

Remember the Mermaid Parade photographers? Or Voice of the Faithful? Or the Ex–Jehovah's Witnesses? All these groups, different as they are in membership, outlook, and goals, share two key characteristics. First, they all started out as latent groups—they had things in common, but the cost and hassle of finding one another was too high. Second, the society they lived in didn't make it easy for them to find one another. In some cases, as with the Mermaid Parade attendees, it was simply because of the old mismatch between effort and outcome. In other cases, though, it was because the institutions best positioned to do the introducing were actively opposed to the goals of the latent group. You could hardly expect the Jehovah's Witnesses or the Catholic Church to spend time or money helping coordinate people who want to criticize them or force them to change their ways of doing business.

Groups like Ex–Jehovah's Witnesses and the Pro-Ana girls

no longer need social support to gather; they all operate under the Coasean floor, where lowered transaction costs have made gathering together so simple that anyone can do it. Recording, searching, and transmitting information, including especially information about ourselves, is something our communications networks are effortlessly good at. The enormous visibility and searchability of social life means that the ability for the like-minded to locate one another, and to assemble and cooperate with one another, now exists independently of social approval or disapproval. The gathering of the Pro-Ana girls isn't a side effect of our social tools, it's an effect of those tools.

When society is changing, we want to know whether the change is good or bad, but that kind of judgment becomes meaningless with transformations this large. It's good that the kids in Belarus now have flash mobs as a tool for opposing political oppression, but for other groups, whether Voice of the Faithful or the passengers demanding better treatment from the airlines, the change looks different depending on where you sit. Loyal Catholics might regard VOTF's demands as a threat to the church they love, and union members may not want the airlines' financial position weakened by the passengers' demands.

Sorting the good from the bad is challenging in part because we're used to social disapproval making it hard for groups to form. Alcoholics Anonymous has more support from society than the Pro-Ana girls, but both groups use the language of self-help to describe what they do. The Pro-Ana movement demonstrates, along with sister movements like Pro-Mia (bulimia) and the Cutters (self-mutilation), that the

definition of self-help has suffered the same blow that journalism has. For much of the twentieth century Alcoholics
Anonymous, the premier self-help organization, set the tone
for social assumptions about self-help: it was a place of devotion and healing, and it promoted a generally approved goal.
The shock of the Pro-Ana movement is that it seems to turn
many of those aspects inside out, helping people remain sick
or become sicker.

The shock turns out to be misplaced: the Pro-Ana movement is in fact a self-help movement, because the content of a
self-help movement is determined by its members. The logic of
self-help is affirmational—a small group bands together to
defend its values against internal and external challenges. When
the small group is a bunch of drunks trying to get sober, against
the norms set by their drinking buddies, then society generally
approves. When the small group is a bunch of teenage girls
trying to get or remain dangerously thin, against the judgment
of their horrified parents and friends, then we disapprove. But
the basic mechanism of mutual support remains the same.

Falling transaction costs benefit all groups, not just groups
we happen to approve of. The thing that kept phenomena like
the Pro-Ana movement from spreading earlier was cost. The
transaction costs of gathering a group of like-minded individuals, especially in an anonymous fashion, has historically
been large, and self-funded and socially approved groups like
AA were the only ones that could take on those costs. Once
the transaction costs fell, however, the difficulties of putting
such groups together disappeared; the potential members of
such a group can now gather and set their own goals without
needing any sort of social sponsorship or approval.

## Three Kinds of Loss

Our new freedoms are not without their problems; it's not a revolution if nobody loses. Improved freedom of assembly is creating three kinds of social loss. The first and most obvious loss is to people whose jobs relied on solving a formerly hard problem. This is the effect felt by media outlets challenged by mass amateurization. The basic problem of copying and distributing information, previously an essential service of the music and newspaper industries among others, is now largely solved thanks to digital networks, undermining the commercial logic of many industries that relied on previous inefficiencies.

Andrew Keen, in *Cult of the Amateur*, describes a firm that ran a $50,000 campaign to solicit user-generated ads. Keen notes that some professional advertising agency therefore missed out on hundreds of thousands of dollars in fees. This loss is obviously a hardship for the ad agency employees, but were they really worth the money in the first place if amateurs working in their spare time can create something the client is satisfied with? The spread of cheap and widely available creative tools is sad for people in the advertising business in the same way that movable type was sad for scribes—the loss from this kind of change is real but limited and is accompanied by a generally beneficial social change.

The second kind of loss will damage current social bargains. Many countries place restrictions on the media in the run-up to elections, but this raises the question of who "the media" is today and what controls should be put on them. Different countries are coming up with different answers—Singapore banned blogging during the last few weeks before a 2005 election but couldn't control Singaporeans blogging overseas; the Thai government forbade blogging on all political matters, to little effect; and the U.S. election commission decided not even to try to apply its media coverage rules to blogging. The provisional and variable nature of these restrictions suggests that the old relations between the media and the state, even where they are broadly supported by the citizenry, are going to be as impossible to sustain as the old definitions for journalism, which is now less a profession than an activity.

The third kind of loss is the most serious. Networked organizations are more resilient as a result of better communications tools and more flexible social structures, but this is as true of terrorist networks or criminal gangs as of Wikipedians or student protesters. This third loss, where the harms are not merely transitional, leads to a hard question: What are we going to do about the negative effects of freedom? It's easy to tell the newspaper people to quit whining, because the writing has been on the wall since the internet became publicly accessible in the early 1990s—their response has been inadequate in part because they waited so long to grapple with the change. It's harder, though, to say what we should be doing about Pro-Ana kids or about newly robust criminal networks.

It used to be hard to get people to assemble and easy for existing groups to fall apart. Now assembling latent groups is simple, and the groups, once assembled, can be quite robust in the face of indifference or even direct opposition from the larger society. (In some cases, that very opposition can *strengthen* the group's cohesion, as with the Pro-Ana girls.)

When it is hard to form groups, both potentially good and bad groups are prevented from forming; when it becomes simple to form groups, we get both the good and bad ones. This is going to force society to shift from simply preventing groups from forming to actively deciding which existing ones to try to oppose, a shift that parallels the publish-then-filter pattern generally.

# CHAPTER 9

# FITTING OUR TOOLS TO A SMALL WORLD

*Large social groups are different from small ones, but we are still understanding all the ways in which that is true. Recent innovations in social tools provide more explicit support for a pattern of social networking called the Small World pattern, which underlies the idea of Six Degrees of Separation.*

Imagine you are seated next to someone on a plane, and after a brief conversation you realize you have a friend or acquaintance in common. At this point you are both required to express surprise at this discovery, and one of you may even make the canonical remark: "What a small world!" After all, what are the chances that the two of you would know someone in common?

The surprising answer is that the chances are actually quite good, for reasons having to do with the structure of social networks. Consider the most basic form of the problem. Given two people drawn at random from a population of six billion, each of you would have to know something like sixty thousand people to have a fifty-fifty chance of knowing someone in common. Even getting to a one-in-ten chance would require that each of you know twenty-five thousand people. Most of us don't know tens of thousands of people, and yet we discover these small-world connections all the time. How is that possible?

The first factor is something called "homophily," or the grouping of like with like. The percentage of the world that rides in airplanes is small, so you are not, by definition, drawn at random from a pool of six billion, you're drawn from a much smaller one. You have at least two other things in common as well (other than both being seated in row nine), and that is your departure and arrival cities, increasing the likelihood that people you know live in a town your seat mate visits and vice versa. The choices you both have made about where to live and work increase the chance that your friends and acquaintances will share a contact.

Now consider your friends. You are probably moderately well connected—neither as social as Paris Hilton nor as reclusive as J. D. Salinger. (This says nothing about you personally—most people fall between those extremes, by definition.) And most of the people you know are (again by definition) in a similarly middling position. It is therefore tempting to assume that everyone is roughly average, but this assumption is wrong (for the same reason that "the average" is meaningless in power law distributions). Assuming that a social network is held together by its average members leads us to underestimate seriously the likelihood of sharing a link with someone we meet. In fact, social networks are held together not by the bulk of people with hundreds of connections but by the few people with tens of thousands.

Consider the list of people you know. You are unlikely to know many recluses, since recluses by definition don't have many contacts. At the other end of the spectrum, you are very likely to know one or more highly connected people, since to be highly connected in the first place, they have to know many people like you. The chance that you are a highly connected person is low, just as it is for everyone, but the chance that you know one is high. And the "knowing someone in common" link—the thing that makes you exclaim "Small world!" with your seat mate—is specifically about that kind of connection. When you are trying to find a link with someone else, you are unlikely to know any given contact of theirs, as we would expect in a sparsely connected environment. But you are very likely to know one of the most connected people they know. It is the presence of these highly connected people that forms the backbone of the social networks.

All this seems like common sense, but it wasn't until 1998 that anyone offered a convincing explanation of the pattern. Prior to that year sociologists understood that social networks somehow manage to be sparsely connected (most people have only a moderate number of connections), but that despite this sparseness the networks are both efficient (any two people are connected together by only a few links—the Six Degrees of Separation pattern) and robust (the loss of a random connection, or even several, doesn't destroy the network). What they didn't understand was how those networks were held together.

In 1998 Duncan Watts and Steve Strogatz published their

research on a pattern they dubbed the "Small World network." Small World networks have two characteristics that, when balanced properly, let messages move through the network effectively. The first is that small groups are densely connected. In a small group the best pattern of communication is that everyone connects with everyone. In a group of friends Alice knows Bob, Carol, Doria, and Eunice, and each of them knows the others. In a cluster of five people there would be ten connections (by Birthday Paradox math), so each person could communicate directly with any of the others. If anyone drops out of the group, temporarily or permanently, none of the other links between people would be disrupted. (This highly connected pattern appears, among other places, in tightly connected clusters of friends using social networks like MySpace and Facebook, or weblogging platforms like LiveJournal and Xanga.)

The second characteristic of Small World networks is that large groups are sparsely connected. A larger collection of people—one that ran from Alice to Zephyr, say—would have many more potential connections. As the size of your network grew, your small group pattern, where everyone connected to everyone, would become first impractical, then unbuildable. By the time you wanted to connect five thousand people (not even the size of a small town) you'd need half a million connections (Birthday Paradox math again). On the other hand, if you let everyone continue to maintain a handful of connections, then as the network grew, any two people pulled at random would have a long chain of links between them, far longer than six links, in fact. Such a network would be unusable, since the people in it would hardly be connected together.

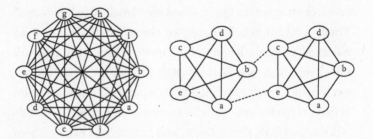

**Figure 9-1:** Two ways of connecting ten people. The left-hand network shows everyone connected to everyone else, which quickly becomes too dense to scale to even moderate numbers of people. The right-hand network keeps people connected but maintains a sparser network.

So what do you do? You adopt both strategies—dense and sparse connections—at different scales. You let the small groups connect tightly, and then you connect the groups. But you can't really connect groups—you connect people within groups. Instead of one loose group of twenty-five, you have five tight groups of five. As long as a couple of people in each small group know a couple of people in other groups, you get the advantages of tight connection at the small scale and loose connection at the large scale. The network will be sparse but efficient and robust.

A Small World network cheats nature by providing a better-than-random trade-off between the number of links required to connect a network, and that network's effectiveness in relaying messages. It occupies a sweet spot between the unbuildable and the unusable, and as a side effect, it is highly resistant to random damage, since the average person doesn't perform a critical function. (By contrast, in a hierarchy almost everyone is critical, since the loss of any one person's connections disrupts communication to everyone connected through that person.) A handful of people are extremely critical to holding the whole network together, because as the network grows large, the existence of a small number of highly connected individuals enables the very trade-off between connectivity and effectiveness that makes the Small World pattern work in the first place.

When you list the participants in a Small World network in rank order by number of connections, the resulting graph approximates a power law distribution: a few people account for a wildly disproportionate amount of the overall connectivity. Malcolm Gladwell, in *The Tipping Point*, calls these people Connectors; they function like ambassadors, creating links

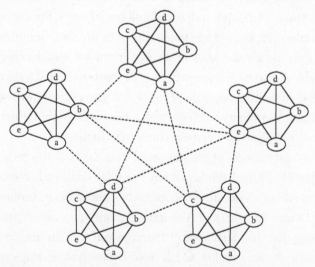

**Figure 9-2:** A network of dense clusters. The network has many fewer connections than it would if everyone were connected to everyone, but it still puts everyone three degrees at most from everyone else. Note that some nodes assume disproportionate importance in holding together the whole.

between disparate populations in larger networks. Without these people large social networks would indeed face the trade-off between impractical and useless. With them everyone is connected to everyone else in six degrees of separation.

So far this is all straight sociology—Watts and Strogatz discovered a pattern common to modern societies, though the connections within and between the smaller groups vary. (Some societies are more tribal than others, with denser local connections and sparser global ones.) What's happening now is that we have tools that both support and extend these patterns. Most Meetup members are members of only one group, but in any large town there are a few people who are members of lots of groups. Meetup is a Small World network, as is MySpace. (Among hundreds of millions of users, the average number of friends is less than sixty, while the median number is five, exactly the kind of disproportion we would expect.) Weblogs also exhibit the pattern—the most connected weblogs are thousands of times more connected than ordinary weblogs are, while ordinary weblogs, with a few readers, are far more likely to be part of a densely connected cluster.

As an example of the way social tools can both rely on and extend the Small World pattern, consider dodgeball, a social networking service designed for mobile phone users, invented by Dennis Crowley and Alex Rainert (both former students of mine). Late last September I found myself with an uncharacteristically free evening, so I decided to stop in at the Magician, a bar on the Lower East Side that some of my friends frequent. Before I arrived, I sent a text message from my phone to dodgeball. The message I sent was simplicity itself: "@magician." The dodgeball service recognized The Magician as a bar (it was previously entered into its database) and recognized me as a registered user. So it was able to send text messages out to my other dodgeball-using friends telling them where I was. They all received messages on their phones saying, "Your friend Clay is at The Magician on Rivington Street."

It also did something more complicated. Because every dodgeball user has a list of friends, dodgeball knows not only that Dennis is a friend of Clay's but that Andy is a friend of Dennis's, who is a friend of Clay's. This is friend-of-a-friend networking (sometimes known as FOAF networking), and it's how social networks like MySpace and Facebook work. But because dodgeball also knows something about the geographic location of its users, and because digital cameras are ubiquitous in the connected crowd that dodgeball targets, it's able to use FOAF networking to broker introductions.

So minutes after I checked in with dodgeball, I got a message back from the service saying, "Andy Krucoff is also at the Magician. You know Andy through Dennis." The message was accompanied by a digital picture of Krucoff. It was small and grainy, but given the uncanny human ability to recognize faces (much of our brain's visual processing apparatus is given over to face recognition), it was enough for me to locate him, even in dim light. Seeing him, I walked up, held out my hand, and said, "I'm Clay. If Dennis were here, he'd introduce us." My meeting Krucoff was simultaneously less social than if no technology had been involved (Dennis, our mutual friend, was nowhere to be seen) and more so (without dodgeball I wouldn't have been able to meet Krucoff at all, even though we were standing ten feet apart). Dodgeball used FOAF networking to take a latent link (in this case, between me and Krucoff running through Dennis) and make it real, or rather it gave me the information I needed to make it real. When I introduced myself, both Krucoff's network and mine got one link denser, and a lot of people I know got one hop closer to him, and vice versa.

The software didn't actually introduce us; it simply provided me the tool to make the introduction myself. Because the number of people you *could* know at any given moment is always a tiny fraction of the people you *do* know, social tools have to help us decide when to actually make a connection. As a result, tools that rely on FOAF networking work better when they augment human social choices rather than trying to replace them. Hundreds of tools build on social networking, from Cyworld (the Korean megasite, with pictorial representations of users) to asmallworld (an intentionally exclusive community for the highly connected and well-off) to Dogster (for dog owners). All make the same underlying assumptions about human links, and all play in some way with the tension between homophily and the desire to meet new people.

Once you've understood this pattern—that a larger network is a sparsely linked group of more densely linked subnetworks—you can see how it could operate at multiple scales. You could tie several few-person networks together into a network of networks. Connections in these larger networks are still between individual people, but now those individuals have become even more critical; in fact, the larger the network is, the more important the highly connected individuals are in holding the overall structure together. Even at seemingly absurd extremes, the pattern holds: random pairs of people from New York City, a pool of millions, are likelier to be connected in a shorter chain than random pairs drawn from the Northeast, and pairs from the Northeast are likelier to be connected in a shorter chain than random pairs from the United States. The layers are arbitrary, but the comparison isn't: because the smaller networks are denser than the larger network of which they are part, the pattern repeats itself at many scales.

Small World networks operate as both amplifiers and filters of information. Because information in the system is passed along by friends and friends of friends (or at least contacts and contacts of contacts), people tend to get information that is also of interest to their friends. The more friends you have who care about a particular piece of information—whether gossip or a job opening or a new song they like—the likelier you are to hear about it as well. The corollary is also true: things that none of your friends or their friends care about are unlikely to get to you. This pair of functions, amplification plus filtering, was at work in the huge MySpace protests in California in 2006. In late March of that year tens of thousands of students in the Los Angeles Unified School District walked out of school and down to City Hall, stopping traffic as they went, as part of a broader protest against a proposed anti-immigration law, HR 4437. Like the Belarusian and Filipino protests, the school boycott assembled quickly, using MySpace pages and mobile phones, and school officials didn't see it coming. The walkout upset the administrators both because it was a threat to their ability to keep order and also because California pays its schools based on average daily attendance, so the student walkout also threatened the schools with a financial penalty. Unlike the old "advertise to everyone but reach a fractional population" model for protests in years past, the Small World network of MySpace and text messaging meant that the message mainly went to students who would already be interested in participating, without becoming public before the event itself.

Small Worlds networks mean that people don't simply connect at random. They connect in clusters, ensuring that they interact with the same people frequently, even in large networks. This in turn reduces the Prisoners' Dilemma and helps create social capital. One reason the phrase "social capital" is so evocative is that it connotes an increase in power, analogous to financial capital. In economic terms, capital is a store of wealth and assets; social capital is that store of behaviors and norms in any large group that lets its members support one another. When sociologists talk about social capital, they often make a distinction between bonding capital and bridging capital. Bonding capital is an increase in the depth of connections and trust within a relatively homogenous group; bridging capital is an increase in connections among relatively heterogeneous groups. Think of the difference by considering the number of people to whom you'd lend money without asking when they'd pay you back. An increase in bridging capital would increase the number of people you'd lend to; an increase in bonding capital would increase the amount of money you'd lend to people already on the list.

One very public illustration of the difference between bridging and bonding capital came in the form of Howard Dean's presidential campaign. At the end of 2003 Dean had the best-funded, best-publicized bid to be the Democratic nominee. Dean was so widely understood to be in the lead that the inevitability of his victory was a broad topic of discussion. Even the people who disputed this inevitability burnished the idea; no one bothered disputing the inevitability of any of the other candidates. And yet Dean's campaign was unsuccessful. It did many of the things successful campaigns do—it got press coverage and raised money and excited people and even got potential voters to aver to campaign workers and pollsters that they would vote for him when the time came. When the time came, however, they didn't. The campaign never succeeded in making Howard Dean the first choice of any group of voters he faced.

The Dean campaign brilliantly conveyed a message to its supporters, particularly its young ones, that their energy and enthusiasm could change the world. Some of this message was conveyed by design, but much of it was a function of people looking for something, finding it in Dean, and then using tools like Meetup and weblogs to organize themselves. The Dean campaign was unequaled in creating bonding capital among its most ardent supporters. They gained a sense of value just from participating; and in the end the participation came to matter more than the goal (a pretty serious weakness for a vote-getting operation). The pleasure in working on the Dean campaign was in knowing that you were on the right side of history; the campaign's brilliant use of social tools to gather the like-minded further fed that feeling. It is natural for a campaign attracting so many eager young people to oversell them on the effect they'll have, when the truth is so rough: you'll work eighty-hour weeks while sleeping on someone's sofa, and in the end your heroic contribution will be a drop in the bucket of what's needed. So a little pep talk now and again can't hurt.

But a campaign can go too far. In this respect, too far is when people believe that believing is enough, without factoring in the difference between the passionate few who run the campaign and the barely interested many who actually vote. Voting, the heart of the matter, is both dull and depressing. Standing around an elementary school cafeteria is not a great way to feel like your energy and excitement are going to change the world, because the math of the voting booth undermines any sense of inevitability: everyone in line not voting for Dean cancels your vote. The Dean campaign had accidentally created a movement for a passionate few rather than a vote-getting operation.

Bonding capital tends to be more exclusive and bridging capital more inclusive. In Small World networks bonding tends to happen within the clusters, while bridging happens between clusters. The Dean campaign was great at doing everything a campaign can do with bonding capital—gathering ardent supporters and raising millions in funds—but getting people to vote for the candidate required bridging capital, reaching out to people outside the charmed inner circle. The Pro-Ana groups also have and use bonding capital well—members are relatively homogenous with regard to age and class and completely so with regard to gender. Meetup relies on (and produces) bonding and bridging capital along a spectrum, depending on the group. A Meetup group for Ping-Pong would produce bridging capital (anyone can join the group, regardless of age, class, or gender), while Black Stay at Home Moms relies on homogeneity as part of the appeal.

The effects of homophily touch every social system; technology doesn't free us from social preferences or prejudices. As danah boyd, the great scholar of social networks (who doesn't capitalize her name), has noted, the populations of MySpace and Facebook, two large social networking sites, mirror divisions in American class structure. Facebook started as a site for college students, so when it opened its virtual doors to high school students as well, it was framed as being for the college-bound, while MySpace remained, in boyd's words, a home for "the kids who are socially ostracized at school because they are geeks, freaks, or queers." Even our modest preferences for bonding can lead to large-scale divisions of this sort.

Perhaps the most significant effect of our new tools, though, lies in the increased leverage they give the most connected people. The tightness of a large social network comes less from increasing the number of connections that the average member of the network can support than from increasing the number of connections that the most connected people can support.

## Bridging Capital, 24/7

Joi Ito is variously an investor, a writer, a hardcore gamer, and a member of the board of scores of companies and nonprofits. His address book contains several thousand names. He is on the road constantly; in 2005 he traveled so much that his average speed was fifty miles per hour for that year. Ito is also an inveterate adopter of new technologies; he tries a remarkable number of social and organizational tools every year and sticks with the ones that make sense to him. One of the tools he adopted a few years ago was internet relay chat, or IRC, an old tool (created in 1988) that creates a real-time chat room called a channel. Everyone using a particular IRC channel can talk to everyone else on that channel. (I use "talk to" here in its colloquial sense of "type quickly to.") Channels on IRC are like instant message or text message conversations, but instead of being people-centric, they are topic-centric. A channel's name gives the users some sense of what the channel's denizens are talking about or at least share in common. Some channels are long-lived—there is a decades-old channel called #hottub (IRC channels always begin with a # sign), mainly dedicated to flirty talk among bored college students the world over. Other IRC channels are transient—a dozen separate IRC channels were set up during the low-speed police chase of O. J. Simpson in 1994, the participants speculating on the result of the chase as it was happening.

In 2004, Ito set up an IRC channel called #joiito, where his friends and contacts could congregate and talk. It was meant to be, as he puts it, "not my place, but a semipublic place, where I could be a host." He used his name both because he is recognized in so many communities (this isn't vanity, just observation—a Web search for "Joi Ito" brings up nearly a million results) and because he wanted to be able to exert some sort of moral suasion over the proceedings. If the IRC channel bore his name, he had a better chance of enforcing civil behavior. The channel quickly grew to have a hundred or so people logged in at the same time. Most of these people were not talking most of the time—they would leave themselves logged in even if they were not paying attention to the channel, or even present at their computers. But their presence ensured relatively regular conversation among the channel's denizens.

One of the regulars, a programmer named Victor Ruiz, wrote a piece of software called jibot (short for "#joiito bot," where a "bot" is an interactive program). Jibot would monitor the channel and answer specially formatted questions, including looking up words in a custom dictionary. One of the regulars brought in a new user, Jeannie Cool, who became an unofficial hostess. She adopted jibot as a social tool by entering in "definitions" tied to the other users' names. Seeing this, Kevin Marks, another regular, modified jibot to make an announcement of this "definition" whenever anyone logged in (a function called heralding). For instance, when the user mmealling shows up, jibot would post the phrase "mmealling is Michael Mealling. He lives in Atlanta, GA." This design, simple as it was, helped move #joiito from a place mainly set up for bonding capital (geeks who knew Joi) into a place that produced bridging capital (people who knew geeks who knew Joi). The jibot, in other words, made #joiito more like a bowling league, which you could join without needing to know most of the members.

The intersection between social networks and electronic networks here is simple in its elements but complex in its results, partly because so many feedback loops are involved. Joi adopted IRC because it was a good way to offer a standing site for interaction among people he knew. The people who gathered there got to know one another better through their interactions, and as with any successful community, new members were invited in. These new members didn't have all the context the original members did, and to meet this need, Ruiz used his technical skills to customize software to create heralding. Critically, the use of jibot as a social tool preceded the heralding function—the reworking of the software was a reflection of behavior, not the other way around. To this day #joiito has about eighty people logged in at any given time. The existence of the channel allows Joi to create a persistent environment where people who know him can meet one another, even if he is not present. Once the channel achieved a kind of social stability, he logged in less and less often; there are people on #joiito around the clock and around the world who do not need a lot of Joi Ito to make it happen. (Dodgeball similarly relies on its users' social capital to help broker introductions without requiring that they be present.) In this way #joiito is an instantiation of Joi's role as a connector; if everyone were to start their own IRC channel, no one would ever talk to anyone else, as they'd all be in their own solo spaces. A channel called #joiito makes social sense—we rely on Joi to provide irreplaceable bridging capital—while one called #jrandomuser wouldn't make sense. Like a bar or a café, #joiito hosts the kinds of informal conversations of which social capital is made; unlike a bar or a café, it costs nothing to run.

Joi's IRC channel is unusual, but the ability of one person or a small group to create this kind of social value is not. Another IRC channel, #winprog, is a hangout for some of the world's most talented Windows programmers. Like many geek hangouts, the channel is a brutal technical meritocracy (rule number one of the channel: "No whining"), but for people serious about Windows programming, #winprog is invaluable, both as a source of information and as a way for serious programmers to get integrated into a community of practice. As a source of both information and camaraderie, #winprog is satisfying and effective for its members.

Similarly, Howard Forums is a Web discussion board founded by Howard Chui, a computer programmer who became obsessive about mobile phones. Chui founded Howard Forums after fielding technical questions from a number of readers to his mobile phone weblog; he reasoned that putting his readers in touch with one another would be easier than trying to answer all their questions himself. The intuition proved correct; less than five years after its founding, the site gets half a billion page-views a year on incredibly detailed subjects, such as customizing specific brands of phones or the merits of various mobile networks. The information produced is so good that engineers at mobile phone companies will sometimes refer customers to it when they have a particularly complex question. Despite the fact that Howard Forums is not an official part of any mobile company, the quality of the technical information there is outstanding, a product of the community's passion for (or obsession with) phones.

Tim O'Reilly, the publisher and conference organizer, founded the conference FOO Camp (Friends of O'Reilly). This conference starts from the invite list—gather a hundred interesting people—and lets them work out the schedule and content of the conference (on a wiki, of course). All these forms suggest that structured aggregation of individual interests and talents can create a kind of value that is hard to replicate with ordinary institutional forms, and impossible to replicate at such low cost.

## It's Not How Many People You Know, It's How Many Kinds

In one of the more evocatively titled papers in the history of social science, "The Social Origins of Good Ideas," Ronald Burt of the University of Chicago detailed his research into the relationship among social capital, social structure, and good ideas. The research method was simple and relatively direct (though interpreting the data wasn't). Burt looked at a major U.S. electronics firm that was undergoing a change in management in 2001, and he got the new management to agree to participate in an experiment. They would ask the managers responsible for the company's supply chain to submit ideas for improving the business and to reveal who else in the firm they'd discussed the idea with, if anyone. This experiment provided a good context for observing the social relations among the company's staff, since the employees responsible for the supply chain were often isolated from the rest of the company. Two of the new managers would then rank the ideas on a scale of 1 to 5; they also had the option to reject an idea outright, if it was "too local in nature, incomprehensible, vague, or too whiny," in the words of one of the senior managers. (Note the resistance to whining, as with #winprog—one of the common requirements to group participation is setting aside purely personal issues in the group setting.)

The essence of Burt's thesis comes down to a linked pair of observations. First, most good ideas came from people who were bridging "structural holes," which is to say people whose immediate social network included employees outside their department. Second, bridging these structural holes was valuable even when other variables, such as rank and age (both of which correlate for higher degrees of social connection), were controlled for. Note that this experiment was a test for bridging capital, not mere sociability—the highest percentage of good ideas came from people whose contacts were outside their own department. On the other hand, managers who were highly connected, but only to others in their department, had ideas that were not ranked as highly. Bridging predicted good ideas; lack of bridging predicted bad ones.

In Burt's analysis, a dense social network of people in the same department (and who were therefore likely to be personally connected to one another) seemed to create an echo-chamber effect. The new managers rejected ideas drawn from this pool with disproportionate frequency, often on the grounds that the ideas were too involved in the minutiae of that particular department and provided no strategic advantage for the company as a whole.

Nor was this experiment a test for intellect. As Burt puts it in the paper:

> People whose networks span structural holes have early access to diverse, often contradictory, information and interpretations which gives them a good competitive advantage in delivering good ideas. People connected to groups beyond their own can expect to find themselves delivering valuable ideas, seeming to be gifted with creativity. This is not creativity born of deep intellectual ability. It is creativity as an import-export business. An idea mundane in one group can be a valuable insight in another.

Burt found that bridging capital puts people at greater risk of having good ideas (his phrase) than do any individual traits. For something like supply chain management, it's easy to see why this might be so—the department handling that function was separated from the rest of the electronics business and was not seen as a core function. The converse was also true; if the proposer of an idea was talking only within his or her own department, the idea was much likelier to be parochial. It seems so simple—just mix people up and sit back and watch the good ideas roll in. There must be a catch.

And there is. Even when the judicious use of social connections increases the proportion of good ideas, most ideas are still bad. It's not enough to find some way to increase the successful ideas. Some way needs to be found to tolerate the failures too.

# CHAPTER 10

# FAILURE FOR FREE

*The logic of publish-then-filter means that new social systems have to tolerate enormous amounts of failure. The only way to uncover and promote the rare successes is to rely, yet again, on social structure supported by social tools.*

The Stay at Home Moms (see Chapter 8) have pressed the generic capabilities of Meetup into service to create a sense of local community that is otherwise hard to arrange in a physically dispersed culture. It's obvious why they would find Meetup valuable. Less obvious but as least as remarkable is *how* that particular group came to be in the first place. Every Meetup group navigates the tension between specificity and size. A Meetup group perfectly fit to an individual (bald fathers of two in Brooklyn who teach at NYU and like bagpipe music) would have exactly one member, while a Meetup that included huge numbers of potential members (parents, or TV watchers, or residents of Atlanta) would provide little in the way of commonality or conversational fodder: "So, you watch

TV too, huh?" The ideal group exists in some equipoise be-tween specific and generic. The Stay at Home Moms group fits that description well enough that it is more popular than all the other parenting groups and one of the most popular Meetup categories overall.

Even accepting that Stay At Home Moms groups exist at some optimum point between size and specificity, there's still a mystery about its formation: How did Meetup know that the group would be as appealing as it was? Most of the people who work at Meetup are overeducated, undermarried urbanites who face a completely different set of problems than do the North Charlotte Stay at Home Moms. How could they have known that SAHM groups would be such a hit?

They didn't. To have predicted such a thing, the employees of Meetup would have needed research about the changing face of American communities, current trends in self-definition of mothers, interactions among suburbanites, and so on; demographics, psychology, sociology. Even if someone had told them that Stay at Home Moms was a good idea for a Meetup group, the staff might have been loath to propose such a thing. Coming from a bunch of single urbanites, it might have seemed patronizing, to say nothing of polarizing. The staff might have become a target for political protest by people upset about the exclusivity implied by the name. Meetup could not have gathered enough information to un-derstand which parenting groups to suggest in the first place, could not have picked a winner even if they'd had all that in-formation, and could not have launched the winner even if they'd been able to pick one, because of the potential nega-tive reaction.

Though it seems funny for a service business, Meetup actually does best not by trying to do things on behalf of its users, but by providing a platform for them to do things for one another. There are hundreds of thousands of Meetup users, and each is presented with many possible Meetups that they could attend. In a midsize city the potential combinations among people interested in Meetup groups are overwhelming. The only sensible way to solve this problem is to turn it over to the users.

The most basic service that Meetup provides is to let its users propose groups and to let other users vote with their feet, like the apocryphal university that lets the students wear useful paths through the grass before it lays any walkways. Most proposed Meetup groups fail because they are too generic, or too specific, or too boring. Most of the rest have only moderate success, leaving only a relative handful of very popular groups, like Stay at Home Moms. This distribution—lots of failure, some modest success, and few extremely popular— is the same pattern (the power law distribution) that we have seen elsewhere. Since failure is normal and significant success rare, Meetup must continually readjust to its current context. It does this by deferring to its users' judgment. The standing question that Meetup poses to its members is "What kind of group is a good idea right now?" Not in the twenty-first century generally, but right now, this month, today. The rise of new groups and the retiring of old ones is not a business decision, it's a by-product of user behavior. Meetup didn't have to establish or even predict the popularity of the Wiccan or LiveJournal groups; nor did it have to predict the time when those groups would be displaced as the most popular. Users are free to propose and pass judgment on groups, and this freedom gives Meetup a paradoxical aspect. First, it is host to thousands of successful groups, groups of between half a dozen and a couple dozen people who are willing to pay Meetup to help them meet regularly, usually monthly, with other people in their community. Second, most of the proposed Meetup groups never take off, or they meet once and never again.

These two facts are not incompatible. Meetup is succeeding not in spite of the failed groups, but *because* of the failed groups. This sounds strange to our ears. Particularly in the world of business, with its Pollyanna-ish attitude toward all public pronouncements, we rarely hear about failure. Meetup's core offer—an invitation for a group of people to get together at a particular place and time—fails with remarkable frequency, as user-proposed groups often don't materialize. Yet Meetup, the company, is doing fine, because the successful groups meet regularly, gain more members, and often spawn new groups in new locations. Meetup is a giant information-processing tool, a kind of market where the groups are the products and where the market expresses its judgment not in cash but in expenditure of energy. Failure is free, high-quality research, offering direct evidence of what works and what doesn't. Groups that people want to join are sorted from groups that people don't want to join, every day. By dispensing with the right to direct what its users try to create, Meetup sheds the costs and distorting effects of managing each individual effort. Trial and error, in a system like Meetup, has both a lower cost and a higher value than in traditional institutions, where failure often comes with some employee's name attached. From a conventional business perspective, Meetup has no quality control, but from another perspective Meetup is *all* quality control. All that's required to take advantage of this sort of market are passionate users and an appetite for repeated public failure.

Meetup shows that with low enough barriers to participation, people are not just willing but eager to join together to try things, even if most of those things end up not working. Meetup is not unusual here. Most pictures posted to Flickr get very few viewers. Most weblogs are abandoned within a year. Most weblog posts get very few readers. On YahooGroups, an enormous collection of mailing lists on topics from macramé to classic TV shows to geopolitics, about half the proposed mailing lists fail to get enough members to be viable. And so on. The power law distribution of many failures and a few remarkable successes is general. Like many of the effects of social tools, this pattern of experimentation appeared first not in services offered to the general public but among software programmers.

## The Global Talent Pool

An interesting effect of digital archiving is that much casual conversation is now captured and stored for posterity, so it is possible to look back in time and find simple messages whose importance becomes obvious only with the passage of time. In the world of software programmers, one of the most important messages ever sent had exactly this casual feel, but it kicked off a revolution. In 1991 a young Finnish programmer named Linus Torvalds posted a note to a discussion group on the topic of operating systems, the basic software that runs computers. In his note he announced his intention to work on a simple and freely licensed system:

> I'm doing a (free) operating system (just a hobby, won't be big and professional like gnu) . . . I'd like to know what features most people would want. Any suggestions are welcome, but I won't promise I'll implement them :-)

The operating system Torvalds proposed that day went on to become Linux, which now runs something like 40 percent of the world's servers (large-scale computers). The existence of Linux has almost single-handedly kept Microsoft from dominating the server market the way it dominates the PC market. Torvalds's brief note contains hints of Linux's future success, hints that are possible to read with the benefit of hindsight. He announced in the first sentence that his new project was to be free. (In a later message he specifically said he intended to use a special software license, the GNU Public License or GPL, to ensure that it stayed free.) The guarantee of freedoms contained in the GPL was critical for encouraging communal involvement; it provided a promise to anyone who wanted to help that their work could not later be taken away. It also ensured that, if Torvalds lost interest in the project, others could pick up where he left off. (He hasn't lost interest, as it turns out, but no one knew what would happen in 1991, nor what will happen in the future.)

Another essential component of Torvalds's original message was that he disavowed world-changing goals. He did not say, "I intend to write software that will prevent Microsoft from monopolizing server operating systems." Instead he made a plausible request—"Help me get this little project started." Linux got to be world-changingly good not by promising to be great, or by bringing paid developers together under the direction of some master plan, but by getting incrementally better, through voluntary contributions, one version at a time.

Finally, Torvalds opened the door, in his first public message, to user participation: "I'd like to know what features most people would want. Any suggestions are welcome, but I won't promise I'll implement them :-)." This kind of openness is the key to any project relying on peer production. The original message got only a few responses. (The population of the internet was only around a million total when Torvalds posted it, less than one-tenth of one percent its size today.) But an early response from someone at the University of Austria indicated some of what was to come.

> I am very interested in this OS. I have already thought
> of writing my own OS, but decided I wouldn't have
> the time to write everything from scratch. But I guess
> I could find the time to help raising a baby OS :-)

The number of people who are willing to start something is smaller, much smaller, than the number of people who are willing to contribute once someone else starts something. This pattern is the same as in the creation of Wikipedia articles, where a simple seven-word entry on asphalt can, through repeated improvement, become a pair of detailed and informative articles. Similarly, enough people have volunteered to help improve Linux that it has gone from a hobby project to an essential piece of digital infrastructure and has also helped propel the idea of collaboratively created (or "open source") software into the world.

Open source software has been one of the great successes of the digital age. The phrase refers to source code, the set of computer instructions written by programmers that then gets turned into software. Because software exists as source code first, anyone distributing software has to decide whether to distribute the source code as well, in order to allow users to read and modify it. The alternate choice, of course, is to distribute only the software itself, without the source code, thus keeping the ability to read and modify the code with the original creators.

Prior to the 1980s, software was something that generally came free with a computer, and much of it was distributed with the source code. As software sales become a business on its own, however, the economic logic shifted, and companies began distributing only the software. One of the first people to recognize this shift was Richard Stallman. In 1980 Stallman was working in an MIT lab that had access to Xerox's first-ever laser printer, the 9700. The lab wanted to modify the printer to send a message to users when their document had finished printing. Xerox, however, had not sent the source code for the 9700, so no one at MIT could make the improvement. Recognizing a broader trend in the industry, Stallman started advocating for free software ("free as in speech," as he puts it). He founded the Free Software Foundation (FSF) in 1983, with a twofold mission. First, he wanted to produce high-quality free software that was compatible with an operating system called Unix. (This project was playfully named GNU, for "GNU's Not Unix.") The second part of the FSF mission was to create a legal framework for ensuring that software stayed free. (This effort led to the GNU Public License, or GPL, which Torvalds was to adopt almost a decade later.)

The year 1983 was a bad time to be arguing for this kind of freedom, as the big computing news was the advent of the personal computer, which was distributed under the "no source code included" model. In the first decade of its existence, FSF seemed to be fighting a losing battle. GPL-licensed software made up an insignificant fraction of the total software in the world, and all of it was used in small and technically adept user communities rather than in the rapidly growing population of home and business users. By the late 1980s it looked like the free software movement was going to be limited to a tiny niche.

That didn't happen, to put it mildly, because the GPL proved useful for holding together much looser groups of collaborators than had ever worked together before, groups like the global tribe now working on Linux. Almost a decade passed between the founding of the FSF and Torvalds's original message. Why did Stallman's vision not spread earlier? And why, after a decade of marginal adoption, did it become a global phenomenon in the 1990s? In that time not much about either software or arguments in favor of freedom had changed. What did change was that programmers had been given a global medium to communicate in. Linux is Exhibit A. When Torvalds announced the effort to build a tiny operating system, he received immediate responses from Austria, Iceland, the United States, Finland, and the U.K., a global collection of potential contributors assembled in twenty-four hours. Within months a simple version of the operating system was up and running, and by then conversations about Linux (as it came to be called) included people in Brazil, Canada, Australia, Germany, and the Netherlands. This had simply been less possible in the 1980s; while there were people online from all those places, they weren't numerous. More is different, and the increased density of people using the internet made the early 1990s a much more fertile time for free software than any previous era.

As Eric Raymond put it in "The Cathedral and the Bazaar," the essay that introduced open source to the world:

> Linux was the first project to make a conscious and successful effort to use the entire world as its talent pool. I don't think it's a coincidence that the gestation period of Linux coincided with the birth of the World Wide Web, and that Linux left its infancy during the same period in 1993–1994 that saw . . . the explosion of mainstream interest in the Internet. Linus was the first person who learned how to play by the new rules that pervasive Internet made possible.

What happened between the founding of the FSF and the creation of Linux, in other words, was a precursor to the things that happened between the two Catholic abuse scandals in Boston, or the stranded planes in 1999 and 2007. Some threshold of transaction cost for group coordination was crossed, and on the far side, a new way of working went from inconceivable to ridiculously easy. All that remains when costs fall is for someone to recognize what has become recently possible. And it was Torvalds who recognized it.

Though the FSF pioneered many of the methods and tools adopted for the creation of Linux, the working methods of Linux were radically different from those of GNU. Stallman is one of the most brilliant programmers ever to have lived, and much of GNU was written by him, or with the help of a few others. Torvalds, by contrast, was crazily promiscuous in soliciting input, though quite judicious in which suggestions he would heed, as he noted in his very first message, "I won't promise I'll implement them." This willingness to listen to a wide group of programmers, coupled with a brutally judgmental meritocracy as to which proposals were worth including, was a radical break with the FSF working method, a break occasioned by the changed transaction costs of gathering the like-minded without a traditional organizational structure. It wasn't just the philosophical commitment to freedom but the scale of the collaboration that made Linux work as software and as a beacon for other open source projects.

## Lowering the Cost of Failure

The Linux project, the most visible open source project in history, has turned the efforts of a distributed group of programmers, contributing their efforts for free, into world-class products. Over the years software produced in this manner has forced significant strategy changes on Microsoft and on other high-tech firms like IBM, Sun, Hewlett-Packard, and

Oracle, all of whom have had to grapple not just with Linux but with other open source programs like Web servers and word processors that are freely available and, more important, freely improvable. But it would be a mistake to assume that because Linux is an open source project, all open source projects are like Linux. In fact, when we look closely at the open source ecosystem, the picture that emerges is characterized more by failure than by success. The largest collection of open source projects in the world is on SourceForge.net, which provides free hosting for software projects. SourceForge boasts more than a hundred thousand open source projects; the most popular of them have been downloaded millions of times cumulatively, and several of them are currently getting more than ten thousand downloads a day. This is the kind of popular attention that the press has focused on when covering open source.

Just beneath these top-performing projects, however, the picture changes. SourceForge ranks hosted projects by order of activity. The projects in the ninety-fifth percentile of activity don't get ten thousand downloads a day; in fact, most haven't gotten even a thousand downloads, ever. These projects are more active than all but 5 percent of what's hosted on SourceForge, and yet they are downloaded less than one tenth of 1 percent as often as the most popular ones.

Projects below the seventy-fifth percentile of activity have no recorded downloads at all. None. Almost three-quarters of proposed open source projects on SourceForge have never gotten to the degree of completeness and utility necessary to garner even a single user. The most popular projects, with millions of users, are in fact so anomalous as to be flukes. (This is, yet again, a rough power law distribution.)

Has the press, then, gotten it wrong about open source? Has it mischaracterized the movement, based on the successes like Linux, when the normal condition of an open source effort is failure? The answer is yes, obviously and measurably yes. The bulk of open source projects fail, and most of the remaining successes are quite modest. But does that mean the threat from open systems generally is overrated and the commercial software industry can breathe easy? Here the answer is no. Open source is a profound threat, not because the open source ecosystem is outsucceeding commercial efforts but because it is outfailing them. Because the open source ecosystem, and by extension open social systems generally, rely on peer production, the work on those systems can be considerably more experimental, at considerably less cost, than any firm can afford. Why? The most important reasons are that open systems lower the cost of failure, they do not create biases in favor of predictable but substandard outcomes, and they make it simpler to integrate the contributions of people who contribute only a single idea.

The overall effect of failure is its likelihood times its cost. Most organizations attempt to reduce the effect of failure by reducing its likelihood. Imagine that you are spearheading an effort for a firm that wants to become more innovative. You are given a list of promising but speculative ideas, and you have to choose some subset of them for investment. You thus have to guess the likelihood of success or failure for each project. The obvious problem is that no one knows for certain what will succeed and what will fail. A less obvious but potentially more significant problem is that the possible value of various projects is unconnected to anything their designers say about them. (Remember that Linus specifically stated that his operating system would be a hobby.) In these circumstances, you will inevitably green-light failures and pass on potential successes. Worse still, more people will remember you saying yes to a failure than saying no to a radical but promising idea. Given this asymmetry, you will be pushed to make safe choices, thus systematically undermining the rationale for trying to be more innovative in the first place.

The open source movement makes neither kind of mistake, because it doesn't have employees, it doesn't make investments, it doesn't even make decisions. It is not an organization, it is an ecosystem, and one that is remarkably tolerant of failure. Open source doesn't reduce the likelihood of failure, it reduces the cost of failure; it essentially gets failure for free. This reversal, where the cost of deciding what to try is higher than the cost of actually trying them, is true of open systems generally. As with the mass amateurization of media, open source relies on the "publish-then-filter" pattern. In traditional organizations, trying anything is expensive, even if just in staff time to discuss the idea, so someone must make some attempt to filter the successes from the failures in advance. In open systems, the cost of trying something is so low that handicapping the likelihood of success is often an unnecessary distraction. Even in a firm committed to experimentation, considerable work goes into reducing the likelihood of failure. This doesn't mean that open source communities don't discuss—on the contrary, they have more discussions than in managed production, because no one is in a position to compel work on a particular project. Open systems, by reducing the cost of failure, enable their participants to fail like crazy, building on the successes as they go.

Cheap failure, valuable as it is on its own, is also a key part of a more complex advantage: the exploration of multiple possibilities. Imagine a vast, unmapped desert with a handful of oases randomly scattered throughout. Traveling through such a place, you would be likely to stick with the first oasis you found, simply because the penalty for leaving it and not finding another oasis would be quite high. You'd like to have several people explore the landscape simultaneously and communicate their findings to one another, but you'd need lots of resources and would have to be able to tolerate vastly different success rates between groups. This metaphorical environment is sometimes called a "fitness landscape"—the idea is that for any problem or goal, there is a vast area of possibilities to explore but few valuable spots within that environment to discover. When a company or indeed any organization finds a strategy that works, the drive to adopt it and stick with it is strong. Even if there is a better strategy out there, finding it can be prohibitively expensive.

For work that relies on newly collapsed transaction costs, however, providing basic resources to the groups exploring the fitness landscape costs little, and the failure of even a sizable number of groups also carries little penalty. Don Tapscott and Anthony Williams tell a story of an almost literal fitness landscape in *Wikinomics*. The mining firm Goldcorp made its proprietary data about a mining site in Ontario public, then challenged outsiders to tell them where to dig next, offering prize money. The participants in the contest suggested more than a hundred possible sites to explore, many of which had not been mined by Goldcorp and many of which yielded new gold. Harnessing the participation of many outsiders was a better way to explore the fitness landscape than relying on internal experts.

Meetup reaps the benefit of this kind of exploration by enlisting its users in finding useful new offerings. By not committing to helping any individual group succeed, and by not directing users in their exploration of possible topics, Meetup has been consistently able to find those groups without needing to predict their existence in advance and without having to bear the cost of experimentation. By creating an enabling service that lets groups set out on their own, Meetup is able to explore a greater section of the fitness landscape, at less cost, than any institution could do by hiring and directing its employees. As with the weblog world operating as an entire ecosystem, services that tolerate failure as a normal case create a kind of value that is simply unreachable by institutions that try to ensure the success of most of their efforts.

The cost of trying things is where Coasean theory about transaction costs and power law distributions of participation intersect. Institutions exist because they lower transaction costs, relative to what a market could support. However, because every institution requires some formal structure to remain coherent, and because this formal structure itself requires resources, there are a considerable number of potentially valuable actions that no institution can afford to undertake. For these actions, the resources invested in trying them will often cost more than the outcome. This in turn means that there are many actions that might pay off but won't be tried, even for innovative firms, because their eventual success is not predictable enough.

It is this gap that distributed exploration takes advantage of: in a world where anyone can try anything, even the risky stuff can be tried eventually. If a large enough population of users is trying things, then the happy accidents have a much higher chance of being discovered.

This presents a conundrum for business. Coasean economics being what they are, a firm cannot try everything. Management overhead is real, and the costs of failures can't simply be laid at the feet of the employees; the firm has to absorb them somehow. As a result, peer production must necessarily go on outside of any firm's ability to either direct or capture all of its value.

This happens in part because the respective costs of filtering versus publishing have reversed. In the traditional world, the cost of publishing anything creates not just an incentive but a requirement to filter the good from the bad in advance. In the open source world, trying something is often cheaper than making a formal decision about whether to try it.

In business, the investment cost of producing anything can create a bias toward accepting the substandard. You have experienced this effect if you have ever sat through a movie you didn't particularly like in order to "get your money's worth." The money is already gone, and whether you continue watching *Rocky XVII* or not won't change that fact. By the time you are sitting in the theater, the only thing you can decide to spend or not spend is your time. Curiously, in that moment many people choose to keep watching the movie they've already decided they don't like, partly as a way to avoid admitting that they've wasted their money.

Because of transaction costs, organizations cannot afford to hire employees who only make one important contribution—they need to hire people who have good ideas day after day. Yet as we know, most people are not so prolific, and in any given field many people have only one or a few good ideas, just as most contributors documenting the Mermaid Parade or Hurricane Katrina contributors contribute only one photo each (the power law distribution again). The institutional response to this imbalance is to ignore the people with only one good contribution; the dictates of 80/20 optimization forces a firm to maximize its output by ignoring casual participants. As a result, many good ideas (or good photos or good music) are simply inaccessible in an institutional framework, because most of the time most institutions have to choose "steady performer" over "brilliant but erratic." It's not that organizations wouldn't like to take advantage of the idea of the occasional participant—it's that they can't. Transaction costs make it too expensive.

In 2005 Nick McGrath, a Microsoft executive in the U.K., had this to say about Linux:

> There is a myth in the market that there are hundreds
> of thousands of people writing code for the Linux
> kernel. This is not the case; the number is hundreds,
> not thousands. If you look at the number of people

who contribute to the kernel tree [the core part of
Linux], you see that a significant amount of the work
is just done by a handful.

If you listen carefully, you can hear McGrath outlining a power
law distribution—only hundreds, not thousands, with the significant work being done by a handful of people.

It's easy to see, from McGrath's point of view, why the
open source model is the wrong way to design an operating
system: when you hire programmers, they drain your resources through everything from salary to health care to free
Cokes in the break room. In that kind of environment, a
programmer who has only one good idea, ever, is a distinctly
bad hire. But employees don't drain Linux's resources, because Linux doesn't have employees, it just has contributors. Microsoft simply cannot afford to take any good idea
wherever it finds it; the transaction costs that come from
being Microsoft see to that. The seemingly obvious advantage of owning the source code carries with it all the overhead of managing that ownership. When Microsoft's
competitors were all commercial firms who faced the same
problems, this overhead was just the cost of doing business,
and bigger firms could rely on economies of scale to compete on overhead costs. The development of Linux, on the
other hand, is not based on the idea of corporate ownership,
which vastly reduces that overhead. Linux can take a good
idea from anyone, and frequently does. It does more than
give Microsoft a new competitor; it changes Microsoft's
competitive environment, as the disadvantages of the insti-

tutional dilemma are no longer uniformly borne by all entrants.

In 2005 Microsoft was desperate to suggest that having an anointed group of professionals, paid to write software, was the only sensible model of development, largely because it had no real alternative. Microsoft operates in a world defined by the 80/20 rule; the cost of pursuing every possible idea is simply too high, so Microsoft must optimize the resources it has. The open source development model, on the other hand, turns the 80/20 rule on its head, asking, "Why forgo the last twenty percent?" If transaction costs are a barrier to taking advantage of the individual with one good idea (and in a commercial context they are), then one possible response is to lower the transaction costs by radically rearranging the relations between the contributors.

The open source movement introduced this way of working, but the pattern of aggregating individual contributions into something more valuable has become general. One example of the expansion into other domains is Groklaw, a site for discussing legal issues related to the digital realm. When the Santa Cruz Organization (SCO), a software publisher, threatened a patent lawsuit against IBM, claiming that IBM's offering Linux to its customers violated SCO's patents, SCO clearly expected that IBM wouldn't want to face either the cost of fighting the suit or the chance of losing and would either pay to license the patents or simply buy SCO outright. Instead, IBM took SCO to court and set about the complex process of uncovering and aggregating what was known about SCO's patents and legal arguments. What SCO hadn't counted on was that Groklaw, a site run by a paralegal named Pamela Jones, would become a kind of third party in the fight. When IBM called SCO's bluff and the threatened suit went forward, Groklaw would post and then explain all the various legal documents being filed. This in turn made Groklaw required reading for everyone interested in the case. The knowledgeable audience that Jones assembled began to post comments related to the case, including, most damningly, comments from former SCO engineers that explicitly contradicted the version of events that SCO was alleging in the trial. Groklaw functioned as a kind of distributed and free friend-of-the-court brief, uncovering material that would have been too difficult and too expensive for IBM to get any other way. The normal course for such a lawsuit would have been that SCO and IBM fought the case in court, while the open source community looked on. What Groklaw did was assemble that community in a way that actually changed the landscape of the case.

## Cooperation as Infrastructure

Emblematic of the dilemmas created by group life, the phrase "free-for-all" does not literally mean free for all but rather chaos. Too much freedom, with too little management, has generally been a recipe for a free-for-all. Now, however, it isn't. With the right kinds of collaborative tools and the right sort of bargain with users, it is possible to get a large group working on a project that is free for all.

McGrath should have been *terrified* that a handful of developers, working alongside a thousand casual contributors, could create an operating system at all, much less one successful enough to compete with Microsoft's commercial offerings. What he misunderstood (or at least publicly misconstrued) was that the imbalance between a few highly active developers and a thousand casual contributors was possible only because Linux had lowered the threshold for finding and integrating good ideas (it reduced the cost of exploring the fitness landscape) in a way that Microsoft simply could not. (Microsoft's Encarta failed to capture user contributions—compare that to Wikipedia.) This problem is not peculiar to Microsoft; as Bill Joy, one of the founders of Sun Microsystems, once put it, "No matter who you are, most of the smart people work for someone else." What the open source model does is to allow those people to work together. This pattern is spreading to other domains; one of the most critical is public health.

Sudden acute respiratory syndrome (SARS), a frequently fatal flu-like disease, first broke out in China in 2002. SARS was, in a way, the first "post-network" virus; enough was known about both the virus and about travel networks to allow airports to prevent travelers from taking the disease with them from continent to continent. These kinds of interdictions kept the disease localized, but they were mere holding actions. What was really needed was an understanding of the disease itself; the race was on to find the genetic sequence to SARS, as a precursor to a vaccine or a cure.

The Chinese had the best chance of sequencing the virus; the threat of SARS was most significant in Asia, and especially in China, which had most of the world's confirmed cases, and China is home to brilliant biologists, with significant expertise in distributed computing. Despite these resources and incentives, however, the solution didn't come from China.

On April 12, Genome Sciences Centre (GSC), a small Canadian lab specializing in the genetics of pathogens, published the genetic sequence of SARS. On the way, they had participated in not just one open network, but several. Almost the entire computational installation of GSC is open source; bioinformatics tools with names like BLAST, Phrap, Phred, and Consed, all running on Linux. GSC checked their work against Genbank, a public database of genetic sequences. They published their findings on their own site (run, naturally, using open source tools) and published the finished sequence to Genbank, for everyone to see. The story is shot through with involvement in various participatory networks.

But if China had the superior intellectual throw-weight and biological research infrastructure, and a bigger incentive than any nation in the world to master the threat, what kept them from winning the race to sequence the virus? One clue came in the form of an interview with Yang Huanming, of the Beijing Genomics Institute, a month after GSC sequenced SARS. Yang said that the barriers in China were not limits on talent or resources, but obstacles to cooperation; the government simply put too many restrictions on sharing either samples of the virus, or on information about it. With considerably fewer resources, GSC outperformed their Chinese counterparts because they'd plugged into so many different cooperative and collaborative networks.

### "Do the People Who Like It Take Care of Each Other?"

In the mid-1990s, at the dawn of the commercial use of the Web, I was in charge of technology for a small Web design firm in Manhattan called Site Specific—there were a dozen of us, working out of the founder's living room. Like the proverbial dog that caught the bus, we landed AT&T as a client. After the ink dried on the contract, AT&T started bringing its engineers around to work with us on programming for the new sites; when we sat down to talk with them, the culture clash was immediate. Site Specific was mostly twenty-somethings (at thirty-one, I was the oldest person in the company), and the AT&T guys (they were all guys) were grizzled veterans who'd been at AT&T longer than most of us had been out of college.

The first real argument we had was around programming languages (a common source of disagreement among techies). AT&T used an industrial-strength language called C++. We used a much simpler language called Perl. The AT&T guys were aghast, and we argued the merits of the two languages, but for them the real sticking point was support. C++ had been invented at AT&T, and they had people paid to support software developers should they run into difficulties. Where, they asked, did we get our commercial support for Perl? We told them we didn't have any, which brought on yet more shocked reactions: We didn't have any support? "We didn't say that," we replied. "We just don't have any *commercial* support. We get our support from the Perl community."

It was as if we'd told them, "We get our Thursdays from a banana"; putting "support" in the same sentence as "community" simply didn't make any sense. Community was touchy-feely stuff; support was something you paid for. We explained that there was a discussion group for Perl programmers, called comp.lang.perl.misc, where the Perl community hung out, asking and answering questions. Commercial support was often slow, we pointed out, while there were people on the Perl discussion group all day and night answering questions. We explained that when newcomers had been around long enough to know what they were doing, they in turn started to answer questions, so although the system wasn't commercial, it was self-sustaining. The AT&T guys didn't believe us. We even showed them how it worked; we thought up a reasonably hard question and posted it to comp.lang.perl.misc. Someone answered it before the meeting with AT&T was over. But not even that could convince them. They didn't care if it worked in fact, because they were already sure it wouldn't work in theory. Support didn't come from evanescent things like an unspoken bargain among a self-assembled community. Support came from solid things, like a contract with a company.

That fight took place a dozen years ago. What's happening today? With the explosion of social tools, the Perl community now has many places to gather, so comp.lang.perl.misc is no longer the epicenter of the community, but it is still a place where people are asking and answering questions, so it's doing fine. AT&T, on the other hand, is not doing as fine. Despite round after round of massive layoffs and alternative strategies, the company shrank to the point of irrelevance, selling itself off to another phone company for $16 billion in 2005, which was only a fifth of its value in 1995, the year it hired us. Perl is a viable programming language today because millions of people woke up today loving Perl and, more important, loving one another in the context of Perl. Members of the community listen to each other's problems and offer answers as a way of taking care of one another. This is not pure altruism; the person who teaches learns twice, the person who answers questions gets an improved reputation in the community, and the overall pattern of distributed and delayed payback—if I take care of you now, someone will take care of me later—is a very practical way of creating the social capital that makes Perl useful in the first place. Between 1995 and 2005 Perl did better as a viable structure than AT&T did, because communal interest turned out to be a better predictor of longevity than commercial structure.

AT&T was right to be skeptical about community; it has not historically been a good guarantor of longevity. The fact that shared interest can now create that longevity is what makes the current change historic. This is the secret of the open source ecosystem and, by extension, of all the large-scale and long-lived forms of sharing, collaborative work, and collective action now being tried. Because anyone can try anything, the projects that fail, fail quickly, but the people working on those projects can migrate just as quickly to the things that are visibly working. Unlike the business landscape, where companies have an incentive to hide both successes (for reasons of competitive advantage) and failures (to forestall any perception of weakness), open source projects advertise their successes and get failure for free. This arrangement allows the successes to become host to a community of sustained interest.

What the open source movement teaches us is that the communal can be at least as durable as the commercial. For any given piece of software, the question "Do the people who like it take care of each other?" turns out to be a better predictor of success than "What's the business model?" As the rest of the world gets access to the tools once reserved for the techies, that pattern is appearing everywhere, and it is changing society as it does.

# CHAPTER 11

# PROMISE, TOOL, BARGAIN

*There is no recipe for the successful use of social tools. Instead, every working system is a mix of social and technological factors.*

Every story in this book relies on a successful fusion of a plausible promise, an effective tool, and an acceptable bargain with the users. The promise is the basic "why" for anyone to join or contribute to a group. The tool helps with the "how"—how will the difficulties of coordination be overcome, or at least be held to manageable levels? And the bargain sets the rules of the road: if you are interested in the promise and adopt the tools, what can you expect, and what will be expected of you? The interaction of promise, tool, and bargain can't be used as a recipe, because the interactions among the different components is too complex. Like the weather, the complex interaction of the various forces makes the results only partially predictable.

The order of promise, tool, and bargain is also the order in which they matter to the success of any given group. Making a promise that enough people believe in is the basic require-ment; the promise creates the basic desire to participate. Then come the tools. After getting the promise right (or right enough), the next hurdle is figuring out which tools will best help people approach the promise together. Wikis make arriv-ing at shared judgment easier than hosting a discussion, while e-mail has the opposite set of characteristics, so getting the tools right matters to the kind of interactions the group will rely on. Then comes the bargain. Tools don't completely deter-mine behavior; different mailing lists have different cultures, for example, and these cultures are a result of an often implicit bargain among the users. One possible bargain for a mailing list is: "We expect politeness of one another, and we rebuke the impolite." Another, very different bargain is: "Anything goes." You can see how these bargains would lead to very dif-ferent cultures, even among groups using the same tools, yet both patterns exist in abundance. A successful bargain among users must be a good fit for both the promise and the tools used. Taken together, these three characteristics are useful for understanding both successes and failures of groups relying on social tools.

The promise is the essential piece, the thing that convinces a potential user to become an actual user. Everyone already has enough to do, every day, and no matter what you think of those choices ("I would never watch that much TV," "Why are they at work at ten p.m.?"), those choices are theirs to make. Any new claim on someone's time must obviously offer some value, but more important, it must offer some value higher than something else she already does, or she won't free up the time. The promise has to hit a sweet spot among several extremes. The original promise of Voice of the Faithful was neither too mundane ("Let's blow off some steam about abusive priests") nor too disrespectful ("Let's demolish the Church"). Instead, its message balanced loyalty with anger— "Keep the faith, change the church." Just right, at least for purposes of recruiting. Similarly, the original message inviting people to work on the Linux operating system was neither too provisional ("Let's try to see if we can come up with something together") nor too sweeping ("Let's create a world-changing operating system"). Instead, Linus's proposal was modest but interesting—a new but small operating system, undertaken principally as a way to learn together. Just right.

The implicit promise of any given group matters more than any explicit one, which is to say that the stated rationale of the group is not necessarily the lived one. The explicit promise of pro-anorexia sites is to be able to get and remain unhealthily thin, but when you read the material posted on those sites, you can see that the actual promise is something more like: "Someone will pay attention to you." Much of the material on these sites is written from the perspectives of girls who have recovered from anorexia: as in other clubs, the pleasure of other people's company is often as important as, and sometimes more important than, the excuse for getting together in the first place.

The problem of getting the promise right is unlike traditional marketing, because most marketing involves selling something that will be made *for* the listeners rather than *by* them. "Buy Cheesy Poofs" is a different message from "Join us, and we will invent Cheesy Poofs together." This second kind of message is more complicated, because of something called the paradox of groups. The paradox is simple—there can be no group without members (obviously), but there can also be no members without a group, because what would they be members of? Single-user tools, from word-processing software to Tetris, have a simple message for the potential user: if you use this, you will find it satisfying or effective or both. With social tools, the group is the user, so you need to convince individuals not just that they will find the group satisfying and effective but that others will find it so as well; no matter how appealing the promise, there's no point in being the only user of a social tool. As a result, users of social tools are making two related judgments: Will I like using this tool or participating in this group? Will enough other people feel as I do to make it take off?

The larger the number of users required, the harder the group is to get going, because the potential users will (rightly) be more skeptical that enough users will join to make it worth their while. (An empty restaurant has the same catch-22 in attracting diners.) There are several strategies for handling this problem. The most obvious one is to make joining easy, in order to make the promise seem within reach. Kate Hanni's Flyers Rights group made the basic action (signing the petition) quite simple and reserved more complicated actions (like calling Congress or talking to the media) for more committed members. Other strategies include creating personal value for the individual users, allowing the social value to manifest only later. Joshua Schachter's service for bookmarking and tagging webpages, called del.icio.us, serves as a personal archive of webpages; the value that accrues from aggregating the group's view of the Web is optional for any given user, but enough people have taken advantage of that value to cause the service to grow dramatically.

Another common strategy is to subdivide the community, in a Small World pattern, so that small but densely connected clusters of people have value even before the service grows large. LiveJournal, the weblogging platform, got a lot of its early growth from clusters of high school students joining at the same time. Though LiveJournal offered more value as it grew bigger (more people to meet, more possible groups to join), it offered enough value to small groups to be able to grow large. (MySpace did something similar during its early growth.) And sometimes good old-fashioned hosting helps bridge the gap, making a promise seem plausible even while users are too few. Part of the promise of Flickr, the photo-sharing platform, was that the public could see your photos. (Flickr made the sharing of photos the default option, though users could turn it off.) Yet the attraction of such photos required an audience, and the logical place to get that audience was from among other Flickr users. Like the proverbial stone soup, the promise would be achieved only if everyone participated, and like the soldiers who convince the townspeople to make the stone soup, the only way to hold the site together before it reached critical mass was through personal charisma. Caterina Fake, one of the founders of Flickr, said she'd learned from the early days that "you have to greet the first ten thousand users personally." When the site was small, she and the other staffers would not just post their own photos but also comment on other users' photos, like a host circulating at a party. This let the early users feel what it would be like to have an appreciative public, even before such a public existed.

Of course, Fake couldn't realistically promise Flickr users that their photos would be admired—most photos are in fact quite dull, on Flickr as elsewhere. What she could tell them was that if they worked to produce admirable photos, they had a chance at finding an audience. The promise of Wikipedia is similarly that you have a chance at having your contributions to an article last, and the promise of weblogs is that you have a chance of finding people who want to read your writing. In the end, because the value of these groups is derived from the participation of the group, the promise is more of a challenge than a guarantee.

## Tools

After the complexity of the question of figuring out the promise of a given group, the question of determining which tools to use seems as if it should be easy. Here again context complicates things. There is no such thing as a generically good tool; there are only tools good for particular jobs. Contrary to the hopes of countless managers, technology is not an infinitely elastic piece of fabric that can be stretched to cover any situation. Instead, a good social tool is like a good woodworking tool—it must be designed to fit the job being done, and it must help people do something they actually want to do. If you designed a better shovel, people would not rush out to dig more ditches.

One surprising ramification of this "goodness of fit" argument is that when you improve the available tools, you expand the number of plausible promises in the world. Linus Torvalds's original promise for Linux seems small in retrospect, but stated baldly—"Let's get a bunch of people all over the world to write incredibly complex software without anyone getting paid"—the proposal would have seemed utterly mad. (Many people treated Linux that way for years, in fact.) Richard Stallman's more managed methods of creating software seemed better than Torvalds's, because up to that point they had been better. In the early 1990s Torvalds's proposal hit the forward edge of what social tools made plausible, and as the tools got better, the size of what was plausible grew. The social tools that the Linux community adopted were like a trellis for vines—they didn't make the growth possible, but they supported and extended that growth in ways that let them defy gravity.

We are living in the middle of a huge increase in the number of available tools: the launch of Twitter, the text-messaging tool, happened *during* the writing of this book. Given this profusion, can we say anything useful about the future social landscape? Yes, but only by switching focus from the individual tools themselves to the kinds of groups the tools are expected to support. Two of the most critical questions are "Does the group need to be small or large?" and "Does it need to be short-lived or long-lived?" Two either/or questions mean four possible combinations; a flash mob is a small short-lived group, while the people contributing to Linux comprise a large and long-lived one, and so on.

The core characteristic of small groups is that their members can interact more tightly with one another, because social density is easier to support in small groups than in large ones (the result of Birthday Paradox math, and part of what drives the Small World pattern). Small groups are thus better conversational environments than large ones and find it easier to engage in convergent thinking, where everyone comes to agree on a single point of view. This is one of the things social tools don't change about group life—small groups are more effective at creating and sustaining both agreement and shared awareness.

The core characteristics of large groups are the inverse. People have to be less tightly connected, on average, to one another. As a result, such groups are better able to produce what James Surowiecki has called "the wisdom of crowds." In his book of that name he identified the ways distributed groups whose members aren't connected can often generate better answers, by pooling their knowledge or intuition without having to come to an agreement. We have many ways of achieving this kind of aggregation, from market pricing mechanisms to voting to the prediction markets Surowiecki champions, but these methods all have two common characteristics: they work better in large groups, and they don't require direct communication as the norm among members. (Indeed, in the case of markets, such communication is often forbidden, on the grounds that small clusters of collaborators can actually pervert the workings of the large system.)

Small and large are relative rather than absolute. In a home, dinner for twelve is (usually) a large group, while in a school a class of a dozen students is small. The coordination issues of dinner are more intense than the coordination issues of discussion. Similarly, a hundred people turning up at a Meetup is a large group, while a hundred people turning out for a political rally is a small one. Whatever the issues of relative size, however, the absolute issue remains—larger groups have looser ties.

Making a promise without having a way to deliver on it isn't plausible, by definition. Tools are tied to the modes of group interaction they need to support. You can see how this is so by imagining switching tools among different groups. The Flyers Rights group and the Egyptian prodemocracy activists both wanted to change the laws of their respective countries. The Flyers Rights group worked deliberately, gathering support over weeks using weblogs and online petition forms. The activists in Cairo used blogs, but some of this activism took place at faster speeds, coordinating in the streets of Cairo using Twitter. Imagine trying to force the Flyers Rights people to use Twitter, while limiting the dozen or so activists trying to save their friend Malek to weblogs. Both groups would have failed. Twitter would have annoyed the people who were happy to sign a petition—the online petition form is a slow-motion but high-visibility tool. Likewise, Alaa Abd El Fattah and his friends could never have coordinated in the streets of Cairo using weblogs—they needed a fast but low-visibility tool like Twitter.

By understanding these two basic constraints of group action—number of people involved and duration of interaction—any given tool, new or familiar, can be analyzed for goodness of fit. And of course a single service can offer more than one tool and thus support more than one form of interaction.

The Windows programmers who hang out on the #winprog chat channel use a tool that supports conversational interaction, but they put their collected group wisdom on a set of webpages, including a FAQ, a list of frequently asked questions, with answers. A FAQ is a social document, representing accumulated wisdom about the commonest questions that arise within a group. The rate at which the FAQ is updated is much slower than the rate at which conversation happens on the chat channel, allowing the community to operate at more than one speed. Similarly, the group that congregates to talk about a particular Wikipedia article may be quite small, while the contributor base of Wikipedia as a whole is enormous, allowing Wikipedia to operate at more than one scale.

Perhaps most important, new tools are not always better. New tools, in fact, start with a huge social disadvantage, which is that most people don't use them, and whenever you have a limited pool from which potential members can be drawn, you limit the social effects. In addition, every social tool is surrounded by an ocean of practice, which helps dictate its use. When the denizens of Bronze, an online discussion group for the TV show *Buffy the Vampire Slayer*, learned that the TV network was no longer going to support their community, they rallied together, raising enough money to commission new software in a new location, like a hermit crab requisitioning a new shell. When they hired a firm to create the new tool, they had one simple request—no major changes. The old tool, which they had gotten used to, was absolutely bare-bones, and they sensed that if the new tool added complicated features, the community would suffer. Instead, they requested (and got) something that looks laughably simple by the standards of more recent software. Their intuition turned out to be correct: the community survived the move from the old location to the new one, which they dubbed Bronze:Beta.

Many of the stories in this book involve the most mundane tools: e-mail lists and discussion groups have been around since the 1970s, and even many of the newer tools, like weblogs and wikis, are already a decade old. The most profound effects of social tools lag their invention by years, because it isn't until they have a critical mass of adopters, adopters who take these tools for granted, that their real effects begin to appear.

## Bargains

The bargain comes last, because it matters only if there is a promise and a set of tools that are already working together. The bargain is also the most complex aspect of a functioning group, in part because it is the least explicit aspect and in part because it is the one the users have the biggest hand in creating, which means it can't be completely determined in advance. The need for a bargain gets back to the most basic issues of group effort—transaction costs. A bargain helps clarify what you can expect of others and what they can expect of you. Imagine you are traveling to a foreign country and are planning to drive while there. Do you want to drive on the right side of the road or the left? The answer, of course, is that right or left is the wrong question—you want to drive on whatever side of the road everyone else does; synchronization with the natives is itself a value. (When in Rome . . .) The same is true for mediated groups; there are many different ways social expectations can be worked out, but as with the rules of the road, what matters is that there is a way and that everyone knows what it is.

On some occasions the bargain with users is simple and elegantly balanced. For example, the basic bargain a wiki offers is that you can edit anyone else's writing, and anyone else can edit yours. Most people who assume wikis will fail because of this freedom underestimate the value of the freedom being extended to everyone.

A few years ago a wiki travel site called WikiTravel.org launched; users were asked to describe various locales in terms that travelers would find useful. Seeing the announcement, I went to take a look. Though the site had been up for only a few days, I found that a basic entry for New York City had already been created. It began:

> "Two for the Glory, Three for the Win" goes the old rallying crying in New York, New York. This former Portuguese colonial capital was a fur and mint-trading mecca in the 18th Century when the Wabash River was a major commercial thoroughfare. The city declined after the growth of railroads allowed travellers land passage from Ontario to Orlando.

This is, of course, all nonsense—not one of those things is true, and the entry went on in that vein for two more paragraphs. Seeing this, I immediately deleted the entry, then looked up the other entries by the same user. It turned out he or she had also created phony entries for Boston ("Boston, known as the 'four-mile-high city,' is perched atop a mountain of more than 20,000 feet in elevation") and Massachusetts ("Massachusetts is technically not a state but a common-whelk"). I deleted those as well, then looked at the history of the various articles that that user had touched. It turned out that he or she had spent the better part of an hour lovingly crafting those three fake entries. I deleted all three in about a minute and a half, and that was that; the prankster never returned, presumably disappointed by the speed with which fake entries could be undone. Wikis take on one of the most basic questions of political philosophy: Who will guard the guardians? Their answer is, everyone. The basic bargain of a wiki means that people who care that the site not be used for that sort of prank have the edge, because it takes far longer to write a fake entry than to fix it.

Other times the bargain is far more one-sided. The bargain for the original flash mobs was that you would show up at the appointed place and time and do as you were told, and in turn you would astonish the people watching you perform these actions in a group. Almost all the power of the original flash mob, in other words, lay with "Bill from New York," and his mobs were designed to mock the hipster willingness to subsume individual judgment in pursuit of confounding the bourgeoisie. The Belarusian flash mobs, by contrast, used the political context to change the bargain considerably (the bargain can never be completely embedded in any given tool), making those who chose to join the mob political actors rather than puppets. Like the choice of the -pedia suffix for Wikipedia, a flash mob is a kind of mental application that both creates and relies on shared awareness among its users. What the Belarusian kids did was to change the mental model of what participation in a flash mob meant.

The essential aspect of the bargain is that the users have to agree to it. It can't be instantiated as a set of contractual rules, because users don't read the fine print. (When was the last time you read through one of those "By clicking here, you agree to . . . " pages on the Web?) Instead, the bargain has to be part of the lived experience of interaction.

Sometimes contracts are an essential part of the bargain, not because of the direct language of the contract but because of what it says about the service. Linus Torvalds offered Linux under the GPL because that was a way of assuring the developers that their work could never be taken away from them. This was an important way he communicated his bona fides years before Linux was valuable enough to appropriate; Torvalds took this step early on to specifically forgo any possibility in the future that he could change his mind and patent or sell Linux. It became valuable precisely because he offered a bargain that limited his future freedom; adoption of the GPL was a serious token of commitment. Wikipedia faced a similar challenge early on. In 2002 the Spanish-language version was growing quickly, but the Spanish users were concerned that Wikipedia might opt for a commercial, ad-driven model. They threatened to take all of their contributions and start an alternate version (a process known as "forking"). This was enough to convince Jimmy Wales to formally forgo any future commercial plans for Wikipedia, and to move the site from Wikipedia.com to Wikipedia.org, in keeping with its nonprofit status. Similarly, he decided to adopt the GNU Free Documentation License for

Wikipedia's content. As with Linus Torvalds's adoption of a GNU license for Linux, the GFDL assured contributors that their contributions would remain freely available, making them likelier to contribute. Years later Wales would remark that Wikipedia was "worth billions," but this claim rewrites history. Given the kind of concern among the various contributors over commercialization one alternate possibility is that the whole thing could have exploded into a dozen different versions, no one of which would be as successful as Wikipedia today. The creation of a formal guarantee that the site's content could never be alienated from its creators helped create the trust necessary for users to commit to it long term, even as it meant forgoing turning Wikipedia into a commercial offering.

Wikipedia also arms its users with ways to help enforce the bargains that make the site work. Wikipedia lists a number of rules for the site, including writing from a neutral point of view and assuming good faith during disagreements. No direct enforcement mechanism is attached to these rules, but users periodically invoke them when they are arguing about the content of an article. This invocation has no formal effect, but it arms the user with a kind of moral suasion that is often enough to settle an argument.

The observations about how tightly the groups need to coordinate with one another also affect how these kinds of social bargains are constructed. As Oliver Wendell Holmes said, "The right to swing my fist ends where the other man's nose begins." In the physical domain cities tend to have more rules that address the vagaries of group life than rural areas do, simply because the number of ways urban life can create social intersections among people is much larger than in rural life. This connection between social density and complexity of bargain is true of technologically mediated groups as well; the more members have to interact with one another, and the more they have to agree to act in concert, the more complex the rules governing their relations have to be. Bargains for sharing can be quite simple while the bargains that have to be worked out for collaboration or collective action are necessarily more complex, because the frequency, complexity, and duration of user interaction are higher. In Holmes's terms, the more integrated a group, the greater the risk of a virtual fist hitting a virtual nose.

The bargain for the group watching Evan Guttman try to retrieve Ivanna's phone was simple—keep watching and talking. By contrast, the various bargains for Flickr can get quite complex. Many of its users are members of groups organized around specific themes ("Telling a Story in 5 Frames," "Street Photography"), and these groups have their own internal expectations: the street photographers don't like staged pictures, while the storytellers don't want single photos. These bargains also involve ongoing negotiation—the basic tension in Flickr groups is a Tragedy of the Commons, where the presence of a potential audience tempts photographers to leave their photos for others to see while not looking at anyone else's. Many of the rules in Flickr groups try to create the kind of mutual coercion that can solve a Tragedy of the Commons, as with this rule for Black and White Maniacs, for takers of black and white photos:

> Post ONE photo, then immediately comment on the
> PREVIOUS TWO photos ... Wait until two more

photos have been posted before posting again. IF
YOU DON'T HAVE TIME TO COMMENT
IMMEDIATELY, PLEASE WAIT TO POST YOUR
PHOTO. It's unfair to expect people to comment on
your work when you're not able to give them that
same courtesy.

The point of such a rule is to make sure that all participants get their photos commented on equally. You may already
have seen the bug in this rule: instead of failing to leave comments (a violation of the letter of the rule), people who want
to take advantage of the assembled audience could leave nearly
worthless comments (a violation of its spirit). Indeed, this is
what happened in Black and White Maniacs, so they added
a clarification:

(**NEW RULE: If you consistently leave one- or two-
word comments like, "nice," "good b+w," "great
catch,". . . . etc. you'll be removed from the group
as well.)

This group prizes evenness of participation among all its
members. This mode of participation is what UCLA anthro-
pologist Alan Page Fiske calls "equality matching," one of four
basic modes of social participation that he has identified. (The
other three are communal sharing, authority ranking, and
market pricing.) The effect of such sharing, in this case of attention, is that the most talented members of the group don't
get much more attention than the least talented; whether this
is good or bad depends on your taste. It would certainly be

possible to have other modes of participation, where an authority determined which photos were worthy of attention, or where members were given play money to set up a market to value the photos, but those groups would be very different in feel from Black and White Maniacs. When you change the bargain, you change the group, even if all the members remain the same.

## Complex Interactions

It's easy enough to keep track of three things. If new forms of group action can be created with a plausible promise, tools fit to the task, and an acceptable bargain, why can't people just put those things on a to-do list and whip up success? Why, in other words, does most proposed group effort fail?

First, because getting each of these elements right is actually quite challenging, while getting all of them right is essential. Second, as with groups themselves, the complexity comes not just from the elements but from their interactions. Remember Larry Sanger's initial mail asking people to contribute to Wikipedia? "Humor me. Go there and add a little article. It will take all of five or ten minutes." He presented it as a favor and as an experiment, and the emphasis was on how easy a wiki would make the process. The simplicity of the tool and the bargain were out front; the promise was little more than "you'll be trying something new and doing me a favor." Compare Jimmy Wales's accounting for Wikipedia's mission now: "Imagine a world in which every single person is given free access to the sum of all human knowledge." The breadth of the explicit promise has risen with the quality of the execution. This explicit promise is different from the implicit one, and it's unlikely that the users making one change to one article once (the commonest case on Wikipedia) are motivated by the stirring language. The implicit promise is simpler: if you help, this will get better.

Tools are similarly complex. Indeed, there is a spectrum of group size even within single communities, since most large groups are sustained through the efforts of a small group embedded within the large one. Because groups often have different subgroups, the bargain is different for different users as well. You can see how complex these interlocking issues become by asking a simple question: Is Wikipedia a community? One obvious answer is: Yes; people work together to create and defend something they clearly love. Another, equally obvious answer is: No, because most contributors add a single item and never interact with anyone. Both answers are right. There is a core Wikipedia community, but it is made up of only a fraction of Wikipedia contributors. The community is drawn from the ranks of the contributors (indeed, moving from reading to contributing is the minimum case for membership), but the community is not the same as the contributors.

The core Wikipedia community could not create Wikipedia alone, because they could not generate enough raw material or take advantage of enough novel points of view. Similarly, the huge but relatively diffuse group of contributors who are not self-consciously part of the community could edit articles, but unless the most committed members of the community defended them, those edits would be destroyed by vandals.

This is not just true of Wikipedia as a whole—it is also true of each individual article, from Asphalt to Zoroaster. Some contributors care about the quality of Wikipedia as a whole, and some care about the quality of any given article, while most just want to fix a typo or add some piece of information they have, and at every level the interaction of these groups holds the whole together. This is what is wrong with a lot of 80/20 optimizations—the belief that truncating the system at the head will optimize its effectiveness; in many cases it actually cuts off a critical piece of the overall ecosystem.

Some parts of a tool are used only by the core contributors. As Fernanda Viegas, the Wikipedia researcher at IBM, has pointed out, Wikipedia has over a dozen separate collections of pages, for functions like the history of specific articles and discussions of them, the administrative functions of Wikipedia itself, and so on. Only one of these collections is for the actual articles; the rest are all about running the site in one way or another. Wikipedia, which looks like a reference work to the average viewer, is in fact a community mainly given over to arguing. The articles are the residue of the argument, being the last thing anyone declined to disagree about. Most of the collections of pages other than the articles, however, are accessed by only the most committed users.

This kind of organization, where small groups form within the framework of a bigger, more diffuse group, is the norm for large collections of people (the Small World pattern at work again), and many large sites are actually designed to help this happen. MySpace, considered as a whole, looks like a tool for a large and long-lived group, but most users don't consider it as a whole. Instead, starting from a "me first, then my friends, then their friends" view of the world, most users see MySpace as a tool for much smaller groups, and this kind of density among groups of friends can make the site a place for much quicker and more tightly organized interactions. The anti–anti-immigration protest in 2006 did not come from MySpace as a whole—News Corp could no more have sponsored such a thing than Meetup could have sponsored the Black Stay at Home Moms in Atlanta. Instead, the tightness of the network meant that users could advertise the walkout to one another, without ever broadcasting a message to the whole site.

Many groups forming today are using software that needs to be customized to a particular group. The groups using Meetup have highly variable rates of success—the Stay at Home Moms groups, which are so popular in the United States, are hardly replicated anywhere else in the world. This kind of customization of a software platform means that the question of promise, tool, and bargain takes place at multiple levels. The basic promise of any Meetup group is that you will be able to meet other people who live near you and share your interests. In addition, each group needs to offer its own specific promise—those of the Ex–Jehovah's Witnesses and the Ping-Pong players will be quite different—and they need to decide what features of the software they will use: Is posting pictures of the Meetups encouraged or forbidden? Can potential members read the message boards, or only actual members? And so on.

Though groups tend first to coalesce around a particular tool, the group is free to adopt additional ones. Jessica

Hammer, a researcher at Columbia University, followed the community that formed around the Web comic "Sluggy Freelance." She found that the community used several different tools, including a discussion group on the Sluggy Freelance website and not one but two mailing lists to coordinate various activities. The group coalesced around one particular site of content, but the internal logic of group cohesion allowed them to expand the number of tools they used. Similarly, though Linux began as a proposal on a usenet discussion group, it has expanded over the years to include multiple mailing lists, websites, and even a custom-created tool for managing the source code itself.

Though a group writing a complex piece of software for itself is rare, customization of the environment to group life is quite common. The Buffy fans who commissioned Bronze: Beta also customized that tool for their community; the Black and White Maniacs on Flickr customized a set of rules for the social bargain they wanted to enforce. Sometimes this sort of customization becomes part of the culture. On alt.folklore .urban, a discussion group for urban folklore, long-time residents used the word "voracity" when they meant "veracity." By doing so consistently, they were able to work newcomers into frenzies of linguistic rectitude; when the newcomers realized they'd been had, they either acquired a newfound respect for the tightness of the community, or they left in a huff. (Needless to say, the regulars viewed either outcome as positive.) This kind of hazing ritual, called "trolling," was not a feature of the social tool alt.folklore.urban was using; it was a norm adopted and sustained by the community.

## All Groups Have Social Dilemmas

In the mid-1960s, the Dutch anarchist group Provo launched its White Bicycle program in Amsterdam. Believing that the political systems of the time had grossly underestimated basic human goodness and had vested too much power in the hands of the state, Provo placed dozens of white bicycles on the streets of Amsterdam, free for all to use. The design of the program was simple: the Provos distributed the bicycles, unlocked and painted white, around the city. You could pick up a bicycle wherever you found it, ride it to your destination, and leave it there for the next person, who would then ride to their next destination and leave it, ad infinitum. In this way a novel communal infrastructure could be made available at low cost, creating a better world for Amsterdam's residents while repudiating the market economy and the "traffic terrorism of a motorized minority." As the Provos put it in their manifesto: "'The white bicycle symbolizes simplicity and healthy living, as opposed to the gaudiness and filth of the authoritarian automobile."

The White Bicycle program would have been just another footnote in the Age of Aquarius, but for one detail: it was an almost instant failure. Within a month all the bicycles had either been stolen or thrown in the canals. Undaunted, many urban visionaries have resurrected the basic idea of the While Bicycles in one way or another. The cumulative results of these experiments are unambiguous: programs that offer unrestricted access to communal bicycles have struggled with theft, and most have ended up collapsing completely, while the communal bike programs that have succeeded have placed restrictions on the use of the bicycles with things like locked sheds and ID cards for checking them in or out. Despite the Provos' optimism, human nature has turned out to be fairly context sensitive; given the opportunity to misbehave, and little penalty for doing so, enough people's behavior becomes antisocial enough to wreck things for everyone. (If only the *Los Angeles Times* had understood this before launching the Wikitorial project.)

The fate of the various White Bicycle programs illustrates, unintentionally but dramatically, a basic truth of social systems: no effort at creating group value can be successful without some form of governance. The groups now adopting social tools form the experimental wing of political philosophy, a place where hard questions of group governance are being worked out. One remarkable aspect of harnessing social value is that social groups tend to be homeostatic, which is to say resistant to external pressures. The classic illustration of individual homeostasis is body temperature. Humans have an internal temperature of 98.6°F, whether they are in the Sahara or in the Arctic. A group, once formed, can achieve homeostasis as well, finding ways to stay together even as the external environment changes.

Pierre Omidyar, a co-founder of eBay, credits the success of his business to trust in the users; he has often said that one of his founding assumptions was that people are basically good. The reality is more complex: eBay may have been founded on a basic trust in human goodness, but within a couple of months after it launched, enough of the transactions were going awry in one way or another that the company had to respond. EBay's solution was to create a reputation system, allowing the buyer and seller in any transaction to publicly report their satisfaction with each other. This system was designed to cast the shadow of the future over both parties, giving each an incentive to maintain or improve their standing on the site; with that addition, eBay became the site we know today. Omidyar was right, with a caveat: people are basically good, when they are in circumstances that reward goodness while restraining impulses to defect. The rewards and restraints can be quite simple and small, but in big groups with relatively anonymous actors, they need to be there or behavior will decay over time.

If a group can last a year it has a good chance of lasting quite a bit longer. The fans of *Buffy the Vampire Slayer* who inhabited the Bronze:Beta bulletin board have remained active long after that series was canceled; Buffy provided the rationale for them to come together, but the value of the community itself became a reason to stay together. Similarly, the Howard Forums community for avid mobile phone users includes a set of off-topic discussions, because the users had been introduced around the topic of mobile phones, but they decided that they liked one another well enough to want to talk about sports, pets, and diets. In extreme cases of homeostasis, the original rationale disappears entirely but the group remains intact. The S-100 Computer Users Group, a Silicon Valley group founded in the late 1970s for users of an early type of personal computer, was still meeting as a social group under that name in the early 1990s, despite the fact that all of the members had long since abandoned their S-100s. The group survived the loss of the founding rationale because they liked one another's company.

This is not to say that all groups become purely social—there is always a tension between the pleasure of community and the rationale for the group's existence (between satisfaction and effectiveness, in other words), and that tension works itself out in different ways. One of my students, Marion Misilim, observed a LiveJournal group in 2002 called "I Love My Boyfriend," populated by girls celebrating the emotion encapsulated in the group's title. During the time Misilim was observing the group, a crisis erupted when one of the group's members had discovered that her beloved had been cheating on her. This discovery had the predictable effect on the relationship, which left the hard question: Now that she no longer loved her boyfriend, could she still be part of the group? Various palliative rationales were offered by the others—she still loved him, even if she despised his actions, or she used to love him and that was enough. These justifications were meant to hold out the possibility of their friend remaining in the group, but to no avail—the wronged girl now loathed her ex-boyfriend, and that was that. The founding promise had not decayed for everyone, as with the S-100 group; it had simply failed for her, and so she left the group.

This is also not to say that groups don't eventually fail. The two longest-lived groups on the internet, the SF-LOVERS and WINE-LOVERS mailing lists, both founded in the early 1970s, each lasted three decades but eventually petered out. The corollary of homeostasis, though, is that most failures happen quickly, usually because of a failure to get one of the big things right.

Many such sites fail the very first test—they offer no plausible promise. After the failure of the *L.A. Times* Wikitorial project, observers discussed whether the problem was with the use of a wiki or with the misbehavior of the users. The answer is, none of the above. The problem with Wikitorial was that the basic offer—"Come help improve the *L.A. Times'* editorials!"—was of little interest in the first place. An editorial is not the sort of content that benefits from group editing, and the promise of working on behalf of the *Times,* a distinctly uncommunal entity, held little appeal except to users who were minded to take advantage of a potential soapbox. Since there was no community that wanted to defend the Wikitorial, the experiment collapsed. Michael Kinsley, then an editor and one of the proponents of the Wikitorial, correctly blamed the failure on the misbehaving users but incorrectly suggested that this condition was unusual. Any experiment of any importance will always have people who want it to fail. Only the systems that have defenses against such users can thrive; the assumption that those users won't appear is an inadequate defense.

If promise were enough, the normal case would be the success, rather than the failure, of social tools. Sometimes the promise is entirely appealing, but the tools are inappropriate. One such example is MoveOn.org, the liberal political organization famous for using the Web to gather support. MoveOn was founded to convince Congress to censure (rather than impeach) President Clinton for lying about his relationship with Monica Lewinsky, and move on to other business; later causes included advocating campaign finance reform and supporting John Kerry's 2004 run for the U.S. presidency. When MoveOn wants to mobilize support, it can e-mail nearly a million users, who in turn send e-mail to Congress.

This hardly sounds like the stuff of failure, but in fact e-mail is the wrong tool for lobbying Congress. Before e-mail, a rule of thumb in Congress was that one handwritten letter from a constituent indicated that something like two thousand voters in that district cared about the same issue. E-mail enormously lowers the transaction cost of sending a message while creating superdistribution, the effortless forwarding of the message from person to person and group to group. The problem with e-mail as a tool is that it is now *too* good—the cost of lobbying Congress by e-mail is so low that an e-mail message has become effectively meaningless. Attempts by Congress to reintroduce some value to the communications—by asking e-mail correspondents to include their mailing address, to make sure that they are a congressperson's constituents—have failed because users can easily cut and paste addresses from the congressperson's district, whether they live there or not. One of the reasons e-mail campaigns continue, despite their near uselessness, is as a public show of force. The individual communications have been denatured, so the battle has moved to public claims of how many mails were sent, which play out in the court of public opinion, not in the halls of Congress. MoveOn, and every other organization that lobbies Congress, would be better served by a less convenient, more expensive tool, one that took real effort to use and so communicated real commitment on the part of the people writing in.

This is happening now. Fans of the TV show *Jericho* were so upset when CBS canceled the show that they started mailing peanuts to CBS in protest using the NutsOnline delivery service. This effort cost the fans real money, so there was no mistaking their commitment, especially not when twenty *tons* of peanuts eventually arrived at CBS. (CBS relented and re-vived the show.) Similarly, antiwar protesters in Michigan and immigrants upset at changes in the handling of U.S. visa rules are protesting by mailing flowers to a Michigan representative and the head of U.S. Immigration Services respectively. Flowers have the dual advantage of signifying respect and being hand-delivered; it's much harder to ignore flowers than e-mails. All these protests have what e-mail lacks, which is proof that the protesters are willing to express their opinion, even at some expense and difficulty.

A similar problem with a change in costs beset the Howard Dean campaign. The first time Dean appeared in the national consciousness was when three hundred people showed up for a Howard Dean Meetup in New York City in early 2003. This level of attendance was unprecedented, and Dean himself took note of it, coming down from Vermont to speak to his support-ers. It seemed like a predictor of great success, but the size of the Dean Meetup was as much a testament to Meetup as to Dean. People were right to be excited about the Dean Meetup but wrong about the reason, because Meetup was founded to lower the coordination costs of real-world gatherings. Prior to Meetup, a turnout of three hundred people would have indi-cated the existence of a huge and latent population of Dean supporters; as with letters to Congress, one individual turning out would have suggested much broader support for Dean. However, because Meetup makes it easier to gather the faith-ful, it confused people into thinking that we were seeing an increase in Dean support, rather than a decrease in transac-tion costs—the 2003 Dean Meetup simply brought out a much larger percentage of Dean supporters than would have shown up previously. (We've seen this sort of effect before, as when written correspondence on letterhead stopped being a sign of a solvent company, thanks to the desktop-publishing revolution.)

Finally, both the promise and the tool can be effective, but the bargain kills the deal. Even Wikipedia's harshest critics concede its popularity and the value in having users report errors. When the managers at Encarta, Microsoft's digital encyclopedia, saw the excitement around Wikipedia, they offered Encarta users the ability to perform a similar service for Encarta. The resulting problem was with neither the promise nor the tool: Wikipedia had shown that people are more than willing to contribute to online reference works, and that the tools are available to do so at low cost and large scale. What doomed the Encarta's effort to minor status was its bargain with users: users had to grant Microsoft permission to "use, copy, distribute, transmit, publicly display, publicly perform, reproduce, edit, modify, translate and reformat your Submission" for a product Microsoft was going to charge money for. This was hardly a bargain at all, as all the power lay with Microsoft, a fact that made the option for user contribution largely irrelevant. Encarta talked the same basic talk, but the details of the bargain, including its tone, undermined the possibility of any large-scale recruitment.

Even when a group offers an obvious promise and provides a stable service, the bargain continues to evolve. In 2004 Flickr was acquired by Yahoo, and the following year Flickr asked its users to switch to using a username and password that were general to Yahoo rather than specific to the Flickr service. It was a technical change, occasioning a minor inconvenience, or so the people running Yahoo thought. Instead, a small but very vocal part of the Flickr population went very publicly berserk, castigating Flickr and threatening to move to another photo service. Little came of the threats; Flickr has millions of users, so it would not have missed the loss of even hundreds, but the fight was never really about the details. What the most vociferous users objected to was the incontrovertible evidence that they were not in control. Although they had contributed the photos that make Flickr what it is, they were not in a position to say no to a unilateral change from Yahoo. The fact that the change was relatively minor didn't matter, because even a minor change exposed the users' relative powerlessness. The incident passed without creating a serious challenge to Flickr, but the amount of public agita is indicative of how seriously users take the implicit bargain, even when (and perhaps especially when) it is not explicitly supported by contract.

A similar revolt from users beset Digg, the user-edited news site, when its staff members found themselves negotiating a bargain with their users that they didn't even know they'd made. The revolt concerned DVDs. All DVDs rely on a secret digital key that is designed to prevent users from copying the contents, and in early 2007 that key was uncovered. People protesting digital restrictions on DVDs began posting the secret number to Digg, and Digg, responding to a request by a DVD industry group, began removing the posts that included it. Digg was not merely within its rights, it was in fact required by law to do so, but the Digg users didn't care. Thousands of them flooded the site with posts containing the key, or instructions on how to find it by searching for "09 F9," and thousands more wrote emails to Kevin Rose, Digg's founder. Their point of view, expressed on a spectrum from polite to livid, was simple: Digg was built on user participation. The users both suggested and rated the stories that appeared on the front page, and in this case what they wanted on the front page was the DVD key. Digg's owners, who thought they were merely complying with a law, came to realize that the users were not casually posting the key—they were engaging in an act of civil disobedience, with Digg as their chosen platform. Faced with exerting unilateral control over their users or living up to their end of the bargain, Digg relented, allowing unlimited posting of the key. As Kevin Rose said, in announcing the new policy:

> [A]fter seeing hundreds of stories and reading thousands of comments, you've made it clear. You'd rather see Digg go down fighting than bow down to a bigger company. We hear you, and effective immediately we won't delete stories or comments containing the code and will deal with whatever the consequences might be. If we lose, then what the hell, at least we died trying.

"At least we died trying." This may be the high-water mark of a commercial firm being held to account by its users. What Rose recognized, and to his credit acted on, was that his business was built not on the software that ran Digg but on the implicit bargain that his users assumed they had with Digg and, by extension, with him. This bargain had nothing to do with the official rules of the site or indeed the legal requirements—the users were intentionally violating both. The bargain was implicit but deeply felt; had Digg's management reneged, the damage to the site's popularity could have been considerable. Once Rose recognized this fact, he took the remarkable step of allowing his company to be a site of collective action by his users, collective action that could have destroyed his business. (Digg survived.)

The Digg revolt is one example of where our social tools are going. Starting with the invention of e-mail, which first functioned to support a conversation in a group, our social tools have been increasingly giving groups the power to co-alesce and act in political arenas. We are seeing these tools progress from coordination into governance, as groups gain enough power and support to be able to demand that they be deferred to. The Digg revolt was one of the broadest examples of this intersection between groups and governance; it will not be the last.

## EPILOGUE

At 2:30 in the afternoon on May 12, 2008, an earthquake hit China's Sichuan province. Hundreds of buildings collapsed, including many schools, leaving nearly 70,000 dead and another 20,000 missing, 350,000 wounded, and 5,000,000 homeless. Word of the quake spread both instantly and globally via social media. The first line of reporting was from Sichuan residents themselves, with messages appearing on QQ (China's largest social network) and on Twitter, even as the ground was still shaking. Within minutes, photos and videos of the quake's effects were being uploaded from mobile phones, with the links being further passed around by e-mail, instant messages, and text messages. The quake was being discussed on QQ and Twitter before it was on any news site; Rory Cellan-Jones of the BBC reported learning of the quake from Twitter. The Wikipedia page for the quake was created within forty minutes to host the now-customary response of sharing links to information about the disaster and its aftermath. Within hours, sites designed to aid the search for missing friends and relatives began popping up, and by the next day, donations from all over the world were being raised on behalf of the survivors.

The speed with which the world became aware of the quake was a function not just of global technological networks, but of its social ones. China and the United States are connected by undersea communications cables, but mere technical connectivity would not have been enough to carry news of the quake as quickly as it did. There is also something we might liken to a social cable running from China to the United States, an invisible bundle of connections between people on both continents. This bundle is made up of all the bonds between the two populations that have built up over the years: the graduate students from China who studied in the United States and returned home, the branch offices of U.S. firms doing business in China; in fact every bit of human contact that makes people want to stay in touch, even when they live far away.

As always, social tools don't create new motivations so much as amplify existing ones. This social cable connects people living in the two countries; when this bundle of connections is supported by social media, the spread of news like the quake is effectively instant, even without mediation by government or official media.

Another reason word of the quake spread so quickly is that it reached a few highly connected individuals, who then passed on what they'd heard to much larger groups. This is Small Worlds network pattern again, where a few well-connected individuals provide the social glue connecting thousands. One such connector was Kaiser Kuo, a Web strategist living in Beijing and engaged in both the U.S. and Chinese Web communities. Kuo was one of the early recipients of news of the quake, and acted as both a translator, from Mandarin to English, and as an amplifier, redistributing the news from China to his contacts around the world. Websites like Global Voices also aggregated news from both amateur and professional sources, serving as a clearinghouse for reports from all over China. The instant and global availability of the news also seems to have pushed the normally cautious Chinese media to begin publishing news of the quake immediately. (In 1976, by contrast, it took the Chinese government several months to admit that a quake of similarly devastating magnitude had even happened.)

The most remarkable aspect of the Sichuan earthquake and social media, though, is likely to be in the future. While it was heartening to see the world's goodwill mobilized in hours, no amount of sympathy or donations could undo the awfulness of the damage. Especially upsetting was the collapse of several hundred schools, killing on the order of five thousand children, a calamity made more awful by China's one child per family policy. These school collapses became part of the story as covered by the official press, in part because the speed and distribution of the reporting from citizens made it impossible to hide. This period of openness lasted for a few days. Then the complaints from the parents started.

Distraught parents claimed that local officials, corrupt and willing to take bribes from construction firms, had turned a blind eye to the substandard construction of schools for years. Had the schools been built according to official standards, they said, the buildings might have remained standing, and thousands of their children would have been saved. This class of protesters was new; they weren't an organized minority, as was true of the Tiananmen Square or Falun Gong uprisings. These were ordinary citizens and grieving parents, radicalized by government failure, and they had a direct line to each other and to the world. This caught the government by surprise; the sudden spotlight of global attention and sympathy after the quake had been welcome (particularly as a distraction from protests over the status of Tibet in the run up to the 2008 Olympics), but the spirit of openness and communication the government had allowed to take hold now made it hard to suddenly reverse course. Unsure what to do, the government waited, but the protests only got more dramatic; these were people who had little left to lose. Before long, images were circulating of local officials literally prostrating themselves in the street before the protesters, begging for forgiveness.

Then, on May 21, the government acted. Official media covering the aftermath of the quake were told to stop reporting on both the collapses and the protests, and were no longer allowed at the site of any of the schools. The experiment in openness was over.

Despite increasingly draconian moves to re-establish control, the Chinese government has not been able to end the protests. A proposed payment to the parents of 60,000 yuan (a bit less than $9,000) was offered only on the condition that they sign a contract agreeing never to bring up the schools issue again; this was widely seen as an insult by the parents, who again made their discontent publicly known. In June, the human rights activist Huang Qi was arrested after offering to aid those who'd lost a child in the collapse, and Liu Shaokun, an employee of the Sichuan school system, was sentenced to a year of "re-education through labor" for posting pictures of the damage on the internet. In July, a group of parents congregating at the mayor's office in the Sichuan city of Mianzhu were dispersed by riot police. Meanwhile, when the government of Hong Kong set up a construction aid fund for Sichuan, they felt compelled to announce that management of the fund would be "highly transparent," a tacit admission of previous opacity in construction funds. The effect of the quake on the local populace, and on their connection to the rest of China and to the world, are still being felt.

It is too soon to know what will happen with the parents' protest—the story is still unfolding, and its ramifications will doubtless be felt for years. It's not too soon to see that social media is changing life in China. In addition to its strong controls over official media, China also maintains what is jokingly called the Great Firewall (a firewall being the name of a protective layer of internet security). The primary goal of this filtering system is to censor overtly political messages coming from media into China from the outside world. What the Great Firewall has not been designed to do is to filter messages from ordinary Chinese citizens heading *out* of China. Services like QQ and Twitter erode the distinction between "media" and "communication," by further fusing personal messages and publicly available forums. On May 12, QQ enabled anyone with a camera-phone to be both a private citizen and a global media outlet. The effects of that change are just beginning to unfold.

The one big lesson from the Sichuan quake is that there is

never just one big lesson. Truly complex events have complex causes and complex ramifications. There are many threads to this story: the effects of social cables of various thickness running between the world's regions, of Small Worlds networks as a natural amplifier of news, of the former audience committing acts of journalism in the quake zone, of the hybridization between professional and amateur media, of the tension between citizen desire for openness and governmental desire for control. All of these are connected pieces of the story, and although they are all patterns we have seen in the world before, their operation during the Sichuan quake was at a scale and level of intensity that dwarfed even the response after the 2005 Indian Ocean tsunami. An event like the quake and its aftermath highlights how ubiquitous, rapid, and global social media has become, but it also accelerates the pace of that change, because once people adopt social media in an unusual situation, they are much likelier to integrate it into their everyday lives.

Increased options for communication in groups don't just mean we will get more of the patterns we already recognize; they also mean we will also get more new kinds of patterns. More is different, even for people who understand that more is different, which explains in part our persistent difficulties in getting technology predictions right.

## Technologies that Matter

When I was growing up, one of the hot debates among my nerd friends was whether we were living in the Atomic Age or the

Space Age. We were certain these were the defining technologies of our era, a certainty inherited from the pages of *Popular Science* and *Popular Mechanics*. The only interesting question was whether limitless energy or the wonders of space flight would transform our world more. We were right to wonder which of two technologies mattered most, but what we didn't know was that we'd picked the wrong two. The most important technologies of the time weren't atomic energy and space flight, they were the transistor and the birth control pill.

The ideal, magazine-cover-ready technology has awe-inspiring engineering and a trivial set of uses. "In the future, we'll have flying cars!" (Really? That sounds great!) "And we'll use them to commute to work!" (Oh.) Atomic energy exemplifies this pattern. Setting up and running a reactor involves complex, dangerous, and pleasingly photogenic tasks, but in the end, nuclear reactors are mainly replacements for older power plants. The characteristics of a transistor were the opposite of the heroic efforts required for nuclear engineering or space flight. What is a transistor but a tiny switch? We've had switches for ages—how big a deal could small be? The very size of the transistor, though, meant that everything in society that touched information would be turned upside down, which has turned out to be a much bigger deal than nuclear energy has been.

When I was a teenager, I remember reading letters to the editor in my local paper, where the grown-ups were arguing about whether to allow students to use calculators. The unspoken worry was that, since calculators had appeared so suddenly, they might disappear just as suddenly. What none of the grown-ups in that conversation understood is that there would never again be a day when we needed to divide two seven-digit numbers on paper. What seemed to them like a provisional new capability was actually a deep and permanent shift, one we students recognized immediately.

Like nuclear power, space flight was similarly removed from any real changes in day to day life. When it came time to imagine commercial space flight, it was presented as if it were plane flight, only higher. There's a wonderful scene in the 1968 movie *2001* (by which time we were all supposed to be traveling to space) where space stewardesses in pink miniskirts welcome the arriving passenger. This is the perfect, media-ready vision of the future—the technology changes, hemlines remain the same, and life goes on much as today, except faster, higher, and shinier. By contrast the birth control pill, like the transistor, seemed to offer only an incremental improvement over existing methods. But by making control of fertility a unilateral and, crucially, a female choice that didn't have to be negotiated case by case, the pill has transformed society in ways far more important than anything ever accomplished by NASA. The movie image of the comely space-stewardesses on the shiny new space station had it backward—since 1968, transportation has not progressed much, but the role of women in society has been transformed.

The transistor and the birth control pill are quite unlike one another, but they do have one thing in common: they are both human-scale inventions that were pulled into society one person at a time, and they mattered more than giant inventions pushed along by massive and sustained effort. They changed society precisely because no one was in control of how the technology was used, or by whom. That is happening again today. A million times a day, someone tries some new social tool; someone in Mozambique gets a mobile phone, someone in Shanghai checks out the Chinese version of Wikipedia, someone in Belarus hears about the flash mob protests, someone in Brazil joins a social networking service.

Much of the world can now use these tools, and within a decade, most of the world will be able to. Mobile phones, which started out as personal versions of ordinary phones, are taking on all the functions necessary to become social tools—digital messaging, the ability to send messages to groups, and, critically, interoperability with the internet, the premier group-forming network (in the sense of both first and best). The global spread of mobile phones has been nothing short of astonishing. In 1994, Greg LeVert, telecommunications engineer, estimated that only half the world had made a phone call. By 2008, there were 3.3 billion mobile phone subscribers, out of a global adult population of less than 5 billion. This increase in scale, both of the underlying social media and of the population that uses it, is still creating surprises because large systems behave differently from small ones.

One fitting name for the way more is different is "the network effect," the name given to networks that become more valuable as people adopt them. Robert Metcalfe, the inventor of the Ethernet networking protocol, gave his name to a law that describes this increase in value. Metcalfe's Law is usually stated this way: "The value of the network grows with the square of its users." When you double the size of the network, you quadruple the number of potential connections. This is Birthday Paradox math, recast as a source of value instead of cost.

Being the only person in the world who can send e-mail isn't a terribly exciting proposition, but once you can send e-mail, every new user means there's someone new you can trade messages with. They spend the money to get online, but as they do, the potential value of your computer rises as well, with the emphasis on "potential." Because of homophily, the value to you that comes from one of your friends plugging into the network is much higher than the value of a random stranger half the world away plugging in, but as we are increasingly seeing with examples like the Sichuan quake, the connections don't all have to be direct to be valuable. Having Kaiser Kuo on Twitter was suddenly more valuable, to many more people, during the quake than it had been before it.

The internet, of course, adds group forming as a possibility, not just person-to-person connections. David Reed, one of the early designers of the internet, has also formulated an eponymous law, which says that the value of group-forming networks actually grows exponentially with the number of users. The logic here is that in a group of four people, there are six ways to set up a two-way conversation (A to B, A to C, etc.), but with a group-forming tools, there can also be four different sets of three-way conversations, or all four in one conversation. With ten people, there are forty-five pairs (Metcalfe's Law), but a thousand possible subgroups (Reed's Law). Reed's Law also relies on the *potential* of communication; the vast majority of possible subgroups will never actually form. The number of potential million-person networks that could theoretically exist on the internet beggars description, but almost none of them actually will, because there's not much a million-person network could do. Most of the action in Reed's Law comes from the formation to human-scale groups—dozens, hundreds, sometimes thousands of people, rather than millions or billions. As with Metcalfe's Law, the growth of the networked population increases the number of potential groups, but the value from Reed's Law grows much faster than Metcalfe's Law, because there are *many* more potential groups than potential pairs.

Metcalfe and Reed's Laws conceive of value to individuals and to groups from all these new options, but what is likely to happen to society as a whole with the spread of ridiculously easy group-forming? The most obvious change is that we are going to get more groups, many more groups, than have ever existed before. Is this a good thing? Is the explosion of new groups pursuing new possibilities with new tools a gain for society? Even accepting that it is painful for many existing organizations and that it produces negative effects as well as positive ones, there are two arguments that suggest that the changes we are living through will be beneficial. The first argument is based on net value, and the second is based on political assumptions.

The net value argument is simple—increased flexibility and power for group action will have more good effects than bad ones, making the current changes, on balance, positive. Examples like the rise of open source software show that new kinds of value are being created all over, and that the good aspects of these new capabilities can outweigh the disadvantages. More recent uses, like the Belarusian kids' flash mobs, the LA school kids' walkout, and the Chinese parents' protests show that as social tools spread, they can be socially and politically relevant as well.

One last comparison with the printing press is instructive

here. As the Abbot of Sponheim correctly saw, the spread of the printed word meant the end of a centuries-old scribal tradition, though once he understood this, he assumed that if scribes were valuable, their loss of livelihood must therefore be bad for society as a whole. The Abbot was laboring under a common economic belief, called the "lump of labor" fallacy. This fallacy is the assumption that there is a certain amount of work in society, a lump of labor, and that any labor-saving device must therefore make society worse off, because people are thrown out of work. In fact, changes like the printing press destroy some kinds of jobs but create others, as well as benefiting a much larger swath of society. Prior to the printing press, much scribal output was given over to simply recopying older material; once the printing press increased the possible supply of books a thousandfold, the price of books fell and the demand rose. The resulting spread of literacy and knowledge benefited society as a whole and led to an explosion in employment for teachers, publishers, scientists, and so on. When old costs are shed, the time and money saved can be applied to new things, things that were unpredictable in the old regime. The profession of Web site designer would have made no more sense to a linotype operator than the profession of typeface designer would have to a scribe.

A subtler weakness to the argument from net value, however, even outside the "lump of labor" fallacy, is that the good and bad changes created by newly flexible groups are incommensurable, which is to say that there is no way of measuring, say, the value of new forms of collaborative action like the kids in Belarus versus the increased resilience of networked terrorist groups. For anyone inclined to see the good effects of the coming changes, a positive value to society can be assured by simply deciding to weigh the benefits more heavily than the disadvantages, while for anyone who believes that the world is going to hell in a handbasket, that conclusion can also be supported by the evidence, simply by deciding that the new bad things are worse or more numerous than the new good things.

The measurement of net value, as appealing as it is, runs aground on this incommensurability, and arguments about whether new forms of sharing or collaboration are, on balance, good or bad reveal more about the speaker than the subject. Net value is a fine tool to use when discussing mere technological improvement—unleaded gasoline is better than leaded, fast trains are better than slow ones, and so on. When there is a real revolution going on, however, net value is useless, since the society before and after the revolution are too different to be readily compared.

The second argument on behalf of new capabilities for groups dispenses with descriptive value, and instead concentrates on political value. In this view, the current changes are good because they increase the freedom of people to say and do as they like. This argument does not suffer from the problems of incommensurability, because an increase in various forms of freedom, and especially in freedom of speech, of the press, and of association, are assumed to be desirable in and of themselves. In this view, the Belarusian kids and Chinese parents have already succeeded in one way, by engaging in political action against the wishes of the government. This does not mean there will be no difficulties associated with our new capabilities—the defenders of freedom have long noted that there are problems peculiar to freer societies. Instead, it assumes that the value of freedom outweighs the problems, not based on a calculation of net value, but because freedom is the right thing to want for society.

The pro-freedom argument does not imply a society with no regulations. Two acts of civil disobedience in the twentieth-century history of the United States demonstrate this. The decisions of much of the population to ignore the constitutional prohibition on alcohol consumption in the 1920s, and the fifty-five-mile-per-hour speed limit in the 1980s, ultimately destroyed those restrictions. The restrictions failed because the cost of enforcement, especially the level of surveillance, was incompatible with a free society. The failure of those regulatory regimes, though, didn't mean that anyone can now drink, or that there is no speed limit. The results of the protests were simply a change to less restrictive regulations. The fundamental tension in the pro-freedom argument is in understanding when freedom can be acceptably limited, within a framework that assumes that the bias should be toward increasing freedom. The basic tenet here is that the unforeseeable effects of freer communication will benefit society, as with the unanticipated rise of an international community of scientists and mathematicians after the invention of the printing press.

Even the pro-freedom argument, though, risks overstating the degree of control we have over the change in group capabilities. To ask the question "Should we allow the spread of these social tools?" presumes that there is something we could do about it were the answer "No." This hypothesis is suspect, precisely because of the kind of changes involved.

Nuclear power is a technology that society can, for the moment, make a decision about. Because of the cost and regulatory strictures implicit in nuclear power in most nations of the world, a country can decide how many nuclear power plants, if any, it wants on its soil. This degree of choice at the national level, though, is tied to the cost of saying "Yes"—billions of dollars of investment and endless vigilance in monitoring its safety. The spread of our social tools is nothing like that— every time someone buys a mobile phone, one of the most routine technological choices possible today, they plug into the grid of social tools and, as we saw after the Sichuan quake, the effects of membership in that grid can be both swift and global. The Chinese parents have both the means and expectation that they can participate in a global conversation, not just because news of the world is rushing in, but because news from the locals is also rushing out.

To put it metaphorically, society's control over nuclear power is like driving a car, with gas, brakes, a reverse gear. We have a good deal of control over both the route and speed with which nuclear power progresses, including the option to simply pull over (as several countries have done by banning the building of new plants). The dramatic improvement in our social tools, by contrast, means that our control over those tools is much more like steering a kayak. We are being pushed rapidly down a route largely determined by the technological environment. We have a small degree of control over the spread of these tools, but that control does not extend to being able to reverse, or even radically alter, the direction we're moving in.

Our principal challenge is not deciding where we want to

go, but rather in staying upright as we go there. The invention of tools that facilitate group formation is less like ordinary technological change, and more like an event, something that has already happened. As a result, the important questions aren't about whether these tools will spread or reshape society, but rather how they do so.

One of the biggest changes in our society is the shift from prevention to reaction, described in relation to the Pro-Ana girls in chapter 8, but fast becoming a more general case. Society simply has less control over what kind of groups can form, and what kind of value they can confer on their members, and this in turn means a loss of prevention as a strategy for reducing harm. Because this change is being brought about by media, there is a good analogy with freedom of speech. In the United States, the First Amendment enjoins the government from limiting the speech of the citizens. There are of course classes of speech that are nevertheless illegal, from shouting "Fire" in a crowded theater to divulging trade secrets, but the legal interpretation of the First Amendment has meant that controls on these illegal classes of speech can't hand the government overly broad powers to restrict speech in advance (a condition called prior restraint) or to create restrictions so broad that the general public feels nervous speaking in public (a condition called a chilling effect). The interaction of these interpretations means that many of the kinds of harm that come from speech are simply allowed to happen, with punishment adjudicated after the fact.

This shift from prevention to reaction is continuing to spread. The Chinese government is losing the ability to shape media coming out of China, a challenge far more serious than censoring inbound information. The French are finding it hard to police hate speech, long illegal but now carried via media increasingly outside their control. The United States can't keep its citizens from gambling online. Governments everywhere are having to increase both surveillance and punishment of pedophiles, now that the pedophiles are able to gather online and trade tips on earning the trust of children. This doesn't mean that these harms will overwhelm us, but it does mean we will have to restructure society, from a strategy of prevention to one of monitoring and reaction, as a side effect of more control of media slipping into the hands of the citizens.

## A Possible Future for Collective Action

Here is a brief account of a scene in a movie theater lobby in a Dallas mall after a showing of Michael Moore's *Sicko*:

> Outside the restroom doors . . . the theater was in chaos. The entire *Sicko* audience had somehow formed an impromptu town hall meeting in front of the ladies' room. I've never seen anything like it. This is Texas goddammit, not France or some liberal college campus. [ . . . ] The talk gradually centered around a core of ten or twelve strangers in a cluster while the rest of us stood around them listening intently to this thing that seemed to be happening out of nowhere. The black gentleman engaged by my redneck in the restroom shouted for everyone's at-

> tention. The conversation stopped instantly as all
> eyes in this group of thirty or forty people were now
> on him. "If we just see this and do nothing about it,"
> he said, "then what's the point? Something has to
> change." There was silence, then the redneck's wife
> started calling for e-mail addresses. Suddenly every-
> one was scribbling down everyone else's e-mail,
> promising to get together and do something . . .
> though no one seemed to know quite what.

Those observations, made by Josh Tyler on the film review site CinemaBlend.com, demonstrate our changed social environment. This group of people, from various backgrounds and brought together only by the accident of a matinee showing at the mall, were able to exchange e-mail, knowing that unlike exchanging addresses or phone numbers, this would let them take the moment of collective inspiration in the lobby and save some part of it for later. These kinds of efforts are unlikely to be long-lived or self-sustaining—no office in D.C., no budget from donations—but the unpredictability of that kind of effort makes it a signal of a kind of commitment that is hard for any ordinary membership organization to produce effectively.

The story of the rapidly coordinated protest by ordinary citizens is one of the most durable stories we have about social media. Since Howard Rheingold's descriptions of the political protest in the Philippines coordinated by text message, we've had countless examples, from the Belarusian flash mob kids to the Latino high school students in LA to the HSBC protesters in the UK. Despite the number of stories about collective action, though, they have one thing in common: they all rely on "stop energy," on an attempt to get some other organization or group to capitulate to the demands of the collected group.

Why is so much collective action focused on protest, with its emphasis on relatively short-term and negative goals? One possible explanation is that it is simply easier to destroy than to create; getting things started in a group takes a lot more energy than trying to stop them. That explanation is hard to support, though, given the fecundity of other kinds of social media. Once you know what to look for, evidence of group creativity is everywhere. I recently came across a site put up by one NickGreat, a member of the Lego figurine-modding community, which consists of pictures of ordinary Lego figurines modified ("modded") with ink and added attachments, so that they resemble various characters in movies or mythology. Viewing the site, you can get some sense of how to mod a Lego figurine yourself so that it will look like characters from a variety of anime cartoons, such as Dragonball Z. This may seem like a trivial example, but triviality is the point—social media is so ubiquitous and cheap that even members of the Lego figurine-modding community find it worthwhile to share. Compare people sharing cute pictures of their cats, sometimes after adding cute captions to the photos. Or the Tax Almanac, a wiki for U.S.-based tax advice. Or the home schooling textbook resale network. Everywhere we look, social media makes creativity not just possible but desirable enough that these examples and millions of others are all out there, with more added every day. Everywhere, that is, except collective action.

Perhaps collective action is more focused on protesting than creating because collective action is simply harder than sharing or collaborating. This at least has the ring of truth about it—collective action is harder to get going because all the participants stand or fall together. The Lukashenko government isn't going to collapse just for some of the flash mob protesters; it's either going to collapse or it's not going to collapse, and everyone in the country will benefit, or not, in the same way. As a result, collective action requires a much higher commitment to the group and the group's shared goals than things like sharing of pictures or even collaborative creation of software.

Even given this difficulty, though, we have examples of people coming together and engaging in collective action that is both long term and creative. The canonical example is a barn raising, where the members of a farming village all turn out to help a neighbor build a new barn, often raised in a single day. A barn raising requires a group; thirty people can raise a barn in a day, but one person can't build the same barn in a month. Barns need groups to do the assembling.

Like open source software and wikis, barn raisings don't involve commercial transactions, and yet they happen. Why would I show up at your farm to help you build your barn when I've got my own work to do? There are two basic answers to that question: either I owe you a favor, or I want you to owe me one. And if either of those things are true of enough individuals, a whole group can enter a state called "reciprocal altruism." With reciprocal altruism, favors are exchanged without formal bookkeeping—if Alice does a favor for Bob, who does a favor for Carol, who does a favor for Doria, and so on, it all comes out in the wash. Instead of each member of the group tracking favors from each other member directly, certain kinds of help simply become social norms.

Barn raisings, though, have one significant limitation—they only work in relatively small communities. Cities don't have anything that feels like a barn raising—certain neighborhoods can achieve something like the necessary social density, but not the city as a whole. And here we have the issue of scale again—the condition of reciprocal altruism is dependent on two things: social density and continuity. Density is required to make reciprocal altruism a strong social norm. Anyone I do a favor for has to know other people who also know me, and this pattern has to be common enough that favors can be passed around the community without formal bookkeeping. When the community is small, that can happen; when it's large, it's easy for free riders to collect favors they simply don't return. The other requirement, continuity, is simply social density in time. Reciprocal altruism requires a kind of communal memory, so that anyone I do a favor for is likely to be around long enough to repay it to me or anyone else in the community.

Small communities with long-time residents have the necessary social density and continuity to build up enough mutual favors to be a fertile ground for reciprocal altruism. Large and transient communities do not—favors "leak" out of the community too quickly. So here's a hypothesis about the near future, based on little more than a hunch and some tantalizing examples: we're about to experience a revolution in collective action, and the driver of that revolution will be new legal structures that will support productive collective action.

All the current examples we have of large-scale, long-lived creativity, like Wikipedia or Linux, are in the realm of intellectual property; Wikipedia and Linux and a million other co-created projects are, in an almost literal way, frozen ideas. What makes most such collaborative efforts work is copyright law, where some form of license is created that allows people to come together and share their work freely, without fear of having that work taken from them later. There are dozens of such licenses, like Richard Stallman's original GPL, currently used by Linux and a host of other collaborative projects, or Creative Commons licenses, which allow the sharing of written work in an analogous manner.

In its twenty-five years of existence, the GPL and its cousins have transformed software development, precisely because they provided assurance to groups of programmers who wanted to pool their efforts, but they are also transforming much of the rest of the software industry as well, because GPL-licensed tools have become such a large part of the ecosystem. In the last ten years, Microsoft has moved from being an implacable foe of Open Source efforts to adopting a position of grudging but genuine accommodation. Similarly, Wikipedia has forced Encyclopedia Britannica to explore opening itself up to both free access for some users, and to taking suggestions from outside contributors. Instead of this kind of change, imagine that Linus Tovalds, originator of Linux, had been limited to protesting Microsoft in order to get a free, high-quality operating system, or that the only way Jimmy Wales and Larry Sanger could get the encyclopedia they wanted was to petition Encyclopedia Britannica to make theirs free. Imagine, in other words, that they had been limited to the tools of protest culture we see in so much of the political sphere. They would have expended far more effort, while accomplishing far less, if anything. What the GPL and related licenses allowed these groups to do was not simply to protest against existing structures, but to compete against them.

This is one of the most interesting differences between social media when it's used for creative and collaborative work, like Wikipedia, compared to its use to coordinate collective action. There is no license for collective action analogous to the GPL, no way a group of people can secure the freedom to work together in ways the government respects. To see what this means in practice, imagine you and a group of five friends walk into a bank, and say, "We're all pitching in together to accomplish something, and we've agreed among ourselves how we want to work together. Please give us a bank account, so our group can start taking, raising, and spending money." You'd be laughed out of the room. The best you could do is have one of you could open a bank account, and add the others as cosigners, and if the original member disappears, the account disappears with her.

Now imagine you and those five friends go off and form a company, and then return to that bank saying, "We've incorporated. Please give us a bank account." The bank would say, "Sign here." An incorporated group can do a number of things an unincorporated group can't, like drafting contracts and bylaws that have legal standing, raising and spending money, hiring and firing employees, and so on. These things are possible in part because incorporation creates both social density and continuity.

The act of incorporation, literally "embodiment," is the way the government recognizes the work of groups, analogous to copyright being the way it recognizes creators. So why don't more groups using social media for long-term goals incorporate? At least part of the answer seems to be that the current corporate structures require things like paper filings, physical headquarters, in-person board meetings, hierarchical management structures, and so on. None of these barriers is fatal by itself, but anything that raises the cost of doing something reduces what gets done. (Coase again.) If there were a structure that allowed for internet-friendly incorporation, we might see an increase in collective action directed at creating and sustaining things, instead of being protest dominated, as it is today.

There are several interesting examples today of just this sort of experimentation. The governor of the state of Vermont recently signed a law that allows for the creation of virtual companies, which allow groups who coordinate mainly or entirely through social media to apply for legal status in Vermont. The new rules governing these virtual companies were designed by David Johnson and his students at New York Law School, with their goal being to allow groups who pool attention and labor and meet online, to have the same kind of legal recognition as companies who pool capital and meet in the real world.

Another approach to the same issue is the Meetup Alliance. The Meetup Alliance grew out of a sense that various Meetup groups in various locales might get value out of associating at a regional, national, or even global level. Meetup provides the infrastructure to allow such a "group of groups" to form, with event listings and discussion areas. In the future, Meetup might also supply fund-raising tools and help creating bylaws. (One unusual decision by Meetup is to allow such alliances to be joined by non-Meetup groups as well, such as mailing lists, Facebook groups, and LiveJournal communities.) It is then up to the groups themselves to decide whether to gather together and what tasks to take on in such an alliance.

While most of these groups don't have overtly political goals—it's hard to imagine the Bellydancing Alliance lobbying Congress—many of the largest alliances are explicitly political, with some dedicated to particular political candidates (echoes of the 2004 Howard Dean Meetup groups) or expressly political goals, like "Claim Democracy," an anticorruption group campaigning for clean government. Most interesting is the possibility of groups with both social and political goals simultaneously. Parents, again, are big adopters of these tools, with alliances like MomsTown and Not Just Moms appearing in the top ten, and it's possible to imagine these groups of groups taking political action around issues of shared interest, possibly even across the same kinds of lines of race and class that characterized the group meeting in the movie lobby after *Sicko*.

These aren't the only such services to experiment with providing better support structure for collective action. Online pledge sites like PledgeBank and ThePoint.com started as "I will if you will" coordination tools, and still tend toward protest (the most successful PledgeBank action to date has been the pledge, by over ten thousand people, to resist the introduction of a national ID card in the UK). However, these sites are exploring ways of helping the groups that form there take constructive action together. There are also purpose-built campaigns, many around environmental issues, like RelightNY, an attempt to get New York City residents to change to compact fluorescent light-bulbs, by engaging them one neighborhood at a time.

Most of the work on supporting collective action around starting or sustaining work is speculative at this point. Most of these efforts didn't exist when I wrote the first draft of this book and neither the Virtual Company nor Meetup Alliances are even a year old at the time of this writing. The appearance of these experiments and others, though, suggests that this is an idea about to undergo a remarkable amount of real-world testing, and the comparison with Open Source licensing suggests that even moderate success may create huge ripple effects for existing institutions (who are, after all, the beneficiaries of the current rules of incorporation). Governments and even companies are accustomed to being the target of protests, so as protests coordinated by social media become normal, their effectiveness will fall. A more remarkable and longer-lived change will be in the offing, though, if people are able to start using these tools to bypass government or commercial entities in favor of taking on problems directly. If this happens, it will be a far bigger challenge to the previous institutional monopoly on large-scale action than anything we have seen to date.

## Taking Change for Granted

In 1501, Aldus Manutius, a Venetian printer, published a translation of Virgil's works. There was nothing particularly unusual about this—by the early 1500s, there were many pub-lishers offering versions of classic texts to an intellectually hungry audience. What was new about Manutius's Virgil was its dimensions—the so-called octavo size was designed to be small enough to fit in a gentleman's saddlebags, so as to make important parts of his library transportable. This was a small revolution, literally and figuratively—small in the sense that the book had shrunk in size and cost, and small in that it was less significant than Gutenberg's original innovation. Yet the octavo size mattered, because it helped spread the written word. By making books cheaper and more portable, Manutius made them more desirable, which in turn meant more copies were produced and more experiments with printing were un-dertaken. In an echo of the salacious nature of many early experiments with content in other media, another of Manu-tius's volumes, *Hypnerotomachia*, was a contemporary novel with erotic passages, a departure from simply translating the classics—and from the contemporary standards of literary propriety. Although the material in *Hypnerotomachia* was cer-tainly less momentous than his editions of Virgil or the Greeks, it helped create a market for new fiction. Manutius's principal insight was to assume, rightly, that the printing press was here to stay. Rather than either lamenting the influence of the press, or continually marveling at its usefulness, he took it on himself to make improvements that seem obvious in retro-spect but which were at the time small revolutions extending the big revolution of movable type.

The lesson from Manutius's life is that the future belongs to those who take the present for granted. One of the reasons many of the stories in this book seem to be populated with young people is that those of us born before 1980 remember a time before any tools supported group communication well. For us, no matter how deeply we immerse ourselves in new kinds of technology, it will always have a certain provisional quality. Our considerable real-world experience usually confers an advantage relative to young people, who are comparative novices in the way the world works. Novices make mistakes from a lack of experience. They overestimate mere fads, seeing revolution everywhere, and they make this kind of error a thousand times before they learn better.

In times of revolution, though, the experienced among us make the opposite mistake. When a real, once-in-a-lifetime change comes along, we are at risk of regarding it as a fad, as with the grown-ups arguing over the pocket calculator in my local paper. What they should have been arguing about instead was how to prepare students to take advantage of the new tools, but they got distracted by assuming that because calculators were new additions to society, they were also provisional ones, when in fact they were new but permanent.

Like Aldus Manutius, young people are taking better advantage of social tools, extending their capabilities in ways that violate old models, not because they know more useful things than we do, but because they know fewer useless things than we do. I'm old enough to know a lot of things, just from life experience. I know that newspapers are where you get your political news and how you look for a job. I know that music comes from stores. I know that if you want to have a conversation with someone, you call them on the phone. I know that complicated things like software or encyclopedias have to be created by professionals. In the last fifteen years, I've had to

unlearn every one of those things and a million others, because those things have stopped being true.

When I spend time thinking about technology, I now
spend more energy on weeding than planting, which is to say
more energy trying to forget the irrelevant than learning about
the new. I've become like the grown-ups arguing in my local
paper about calculators; in the same way it took them a long
time to realize that calculators were never going away, those
of us old enough to remember a time before social tools became widely available are constantly playing catch up. Meanwhile, my students, many of whom are fifteen years younger
than me, don't have to unlearn the thousands of things I do,
because they never had to learn them in the first place.

The advantage of youth, however, is relative, not absolute—
just as everyone eventually came to treat the calculator as a
ubiquitous and invisible tool, we are all coming to take our
social tools for granted as well. The ability of people to share,
cooperate, and act together is being improved dramatically by
our social tools. As everyone from working biologists to angry
air passengers starts to adopt those tools, it is leading to an
epochal change.

## ACKNOWLEDGMENTS

No one knows better than I that a book is the work of a group, and that the thanks I can offer will be both incomplete and inadequate to the debt.

First, I must thank Red Burns for recruiting me to the Interactive Telecommunications Program at NYU; she has created not just a wonderful environment for teaching but for thinking. Dan O'Sullivan, the associate director, and my colleagues Tom Igoe and Nancy Hechinger offered vital comments and support. I must also thank many of my former students, who have always asked sharp questions and pushed for clear answers. Jessica Hammer, in particular, was an ideal co-conspirator, and watching Dennis Crowley and Alex Rainert build Dodgeball was a revelation. Other students who have been especially incisive on social topics are Mouna Andraos, Jake Barton, Michelle Chang, John Geraci, Elizabeth Goodman, Christina Goodness, Sam Howard-Spink, James Robinson, Matty Sallin, Nick Sears, Mike Sharon, and Shawn van Every. Alicia Cervini's careful reading has improved both the ideas and their expression from the first draft.

My field has a tradition of thinking out loud. Chris Anderson, Andrew Blau, Stewart Brand, Lili Cheng, Esther Dyson, Hal Levin, Bob Metcalfe, Jerry Michalski, Richard O'Neill, Tim O'Reilly, Peter Schwartz, Andrew Stolli, and Kevin Werbach all provided both observations and public platforms for the development of this work. Articles written for Chris Anderson for *Wired* and Thomas Stewart for *Harvard Business Review* did likewise.

Long-running conversations with many colleagues have provided material and insights for this book. This list, of all, is both long and incomplete. Researchers, academic and corporate, who have provided critical insights include Yochai Benkler, danah boyd, Elizabeth Churchill, Susan Crawford, Richard Hackman, David Johnson, Valdis Krebs, Frank Lantz, Beth Noveck, Paul Resnick, Linda Stone, Jon Udell, Fernanda Viegas, Martin Wattenberg, and Ethan Zuckerman. Other writers and thinkers with whom I've had illuminating conversations include Cate Corcoran, Cory Doctorow, Ze Frank, Dan Gillmor, Adam Greenfield, Bruno Guissani, Jeff Howe, David Isenberg, Joi Ito, Xeni Jardin, Steven Johnson, Matt Jones, Quinn Norton, Danny O'Brien, Kevin Slavin, Alice Taylor, and David Weinberger. Business colleagues who have provided both observations and theories of mediated social dynamics include Marko Ahtisaari, Stewart Butterfield, Tom Coates, Rael Dornfest, Greg Elin, Caterina Fake, Seth Goldstein, Marc Hedlund, Scott Heiferman, Tom Hennes, J. C. Herz, Sara Horowitz, Ray Ozzie, and Joshua Schachter. My fellow bloggers at Many-to-Many have been supremely good company: Ross Mayfield, Seb Paquet, and especially Elizabeth Lane Lawley.

My agent John Brockman helped me clarify what I wanted to say, Vanessa Mobley and Scott Moyers of Penguin and Kate Monahan of ITP helped me say it, and Janet Biehl provided a marvelous edit.

Finally, of course, is Almaz Zelleke, my wonderful wife, who looked across the dining room table one day and said, "Time for you to write a book," and who has been unerringly patient with the process and an invaluable reader of the product from then until now.

Thank you all.

# BIBLIOGRAPHY

Our communications networks now overlap, with personal and public communications taking place in the same medium. Much of the material covered in the book is available in written form, in the sense of being on a webpage somewhere, but isn't contained in traditional published literature. As a result, readers interested in additional material on the subjects of flash mobs or photo sharing, say, will find themselves looking at a mix of personal and professional media.

This section is a mix of bibliographic references, notes, and pointers to additional material on the Web. Although the best way to find much of the internet-native material in the book is simply to use a search engine, I have tried to include the URLs of sites I mention here, where those URLs are not included in the body of the text. (I have omitted the http:// that is at the beginning of valid URLs, both here and in the body of the book, because most browsers fill that in automatically.) In addition to calling out specific resources, Wikipedia is an especially good guide to many of the general topics covered in this book, precisely because everyone who contributes to

Wikipedia is comfortable using social tools. Wikipedia is useful both for a basic gloss on the related concepts and because at the bottom of most Wikipedia articles (and all materially complete ones) there is a list of additional resources.

One disturbing feature of Web media is their potential evanescence. Because many sites are labors of love (for reasons discussed in the book), there is no guarantee that the materials will last for years, much less decades. Many organizations are working on long-term solutions to this problem; the most fully realized effort is Brewster Kahle's Internet Archive, at archive.org. Among the services hosted at the Internet Archive is the Wayback Machine, which contains snapshots of an enormous number of websites taken over a period of years. For instance, a search of the Wayback Machine for material relating to the story of Ivanna's phone produces a list of archived copies of Evan's website, available at the rather lengthy URL web.archive.org/web/*/evanwashere.com/ StolenSidekick (the * is part of the URL). The Wayback Machine has only a fraction of the material produced for the Web since the early 1990s, but its collection is far larger and more general than any other publicly available resource.

CHAPTER 1: IT TAKES A VILLAGE TO FIND A PHONE

Page 1: **Ivanna's phone** Most of the material on the loss and recovery of Ivanna's phone comes from Evan Guttmann's posts at evanwashere.com/StolenSidekick. That page includes his own account of the events, as well as links to both the bulletin board conversations about the phone, and links to other media coverage of the events. The most relevant Web search is "stolen sidekick," as that is what Evan titled his page and was the phrase most often used by others writing about it. (Search for the phrase in quotes, so you are searching for the phrase, rather than simply searching for pages that happen to contain both words.)

Page 7: *We the Media: Grassroots Journalism By the People, For the People*, O'Reilly Media (2004). Dan Gillmor, a career journalist, also founded the Center for Citizen Media (www.citmedia.org) in 2004.

Page 17: **an architecture of participation** Tim O'Reilly's description of his phrase "architecture of participation" is at www.oreillynet.com/pub/a/oreilly/tim/articles/architecture_of_participation.html.

Page 18: **a plausible promise** Eric Raymond's seminal 1997 essay on open source software, "The Cathedral and the Bazaar," is at catb.org/~esr/writings/cathedral-bazaar/. Raymond's writings on software and other topics are at www.catb.org/~esr/writings/.

Page 22: *Within the Context of No Context*, George W. S. Trow, Atlantic Monthly Press (1997).

## CHAPTER 2: SHARING ANCHORS COMMUNITY

Page 25: **Birthday Paradox** Wikipedia contains a good general guide to the Birthday Paradox, at en.wikipedia.org/wiki/Birthday_paradox. (As always, Wikipedia also contains links at the bottom of the article to additional materials on the subject.) An alternate formulation of the same math is expressed as "Metcalfe's law." Robert Metcalfe, inventor of a core networking technology called Ethernet, proposed that "the value of a network rises with the square of its members," which is to say that when you double the size of a network, its value quadruples, because so many new links become possible. Metcalfe's law isn't true in any literal sense, because not all links are created equal—being able to contact your friends matters more than being able to contact someone you've never heard of who lives on the other side of the world. However, it does capture a basic truth, which is that communications networks grow disproportionately valuable as they grow large.

Page 28: **"More Is Different,"** *Science* 177 (4047), August 4, 1972, pp. 393–96. Philip Anderson's article was a direct attack on the reductionist strategy of science, in which systems are reduced to their simplest elements and those elements become the object of study. Most of the work cited in this article presumes that aggregations of people exhibit properties not reducible to individual behaviors, the core point of "More Is Different."

Page 29: *The Mythical Man-Month: Essays on Software Engineering*, Frederick P. Brooks, Jr., Addison-Wesley (1975). Though this book is filled with interesting general observations about the design of software, the observation that adding more programmers to a late project makes it later is by far Brooks's most famous observation, and the one with the broadest applicability beyond the world of software engineering.

Page 30: **"The Nature of the Firm,"** R. H. Coase, *Economica* 4(16), November 1937, pp. 386–405, also at www.cerna.ensmp.fr/Enseignement/CoursEcoIndus/SupportsdeCours/COASE.pdf. This short work comes from an era when economics papers were readable, and the clarity and elegance of the argument makes the paper relevant even today.

Page 31: *Leading Teams: Setting the Stage for Great Performances*, J. Richard Hackman, Harvard Business School Press (2002).

Page 31: **the Mermaid Parade** Everything you might want to know about the Mermaid Parade is at www.coneyisland.com/mermaid.shtml. Images of the parade are at www.flickr.com/photos/tags/mermaidparade, and at URLs with dates added, as with www.flickr.com/photos/tags/mermaidparade2007.

Page 33: **tagging** was christened by Joshua Schachter, the inventor of del.icio.us, a service for listing and describing webpages, and later recommended as the method for Flickr. The surprise with tagging is that the aggregate judgment of the users provides a useful categorization of webpages without requiring any professional catalogers. I make this argument in considerably more detail in "Ontology is Overrated" at shirky.com/writings/ontology_overrated.html.

Pages 34–38: **the London bombings, Indian Ocean tsunami,** and **Thai coup** After awareness of any major crisis or catastrophe, a Wikipedia page will be created and populated almost instantly and will receive multiple edits in a short period as the details become known. This was not just the pattern with the examples quoted here; it is the normal case for any newsworthy event that affects a large number of people. Wikipedia in turn acts not just as a repository of information; it also acts as a repository for pointers to other resources on the same topic, as with pictures on Flickr or relevant blog posts like those from gnarlykitty.

Page 40: *The Visible Hand: The Managerial Revolution in American Business,* Alfred D. Chandler, Jr., Harvard University Press (1977). The connection between the org chart discussed by Chandler and networked forms of organization was made by David Weinberger in his book, *Everything Is Miscellaneous: The Power of the New Digital Disorder,* Times Books (2007), which describes the radical shift we can make in describing the world once we can do so using digital tools.

Page 47: **cooperation** The literature on cooperation is vast and somewhat confusing. Cooperation per se does not need explaining; birds and bees do it, as the song goes. The much harder question is how we came to engage in so much cooperation with people we are not related to. There is still no good cross-disciplinary explanation for this phenomenon; there are proposed explanations in economics, biology, psychology, and sociology, but while many of these explanations overlap, they have not yet been synthesized.

In economics, *The Origin of Wealth: Evolution, Complexity, and the Radical Remaking of Economics,* by Eric D. Beinhocker, Harvard Business School Press (2006) provides a literature review of economic work on cooperation and its effects. *Small Groups as Complex Systems: Formation, Coordination, Development, and Adaptation,* by Holly Arrow, Joseph E. McGrath, and Jennifer L. Berdahl, Sage (2000) provides a good review of work on small group dynamics, and *Why Humans Cooperate: A Cultural and Evolutionary Explanation,* by Natalie Henrich and Joseph Henrich, Oxford University Press (2007) provides a one for larger groups. Howard Rheingold, whose *The Virtual Community: Homesteading on the Electronic Frontier,* Basic Books (1993) was a critical early work on online community, is working on a multiyear study of cooperation in collaboration (www.cooperationcommons.com) with the Institute for the Future.

Page 51: "**The Tragedy of the Commons,**" Science 162 (3859), December 13, 1968, pp. 682–83. Garrett Hardin was a biologist, and the tragedy of the commons formulation often appears in discussions about natural resources. (There's an online version at www.garretthardinsociety.org/articles/art_tragedy_of_the_commons. html).

A more mathematically rigorous view of the same problem appears in Mancur Olson's *The Logic of Collective Action: Public Goods and the Theory of Groups,* Harvard University Press (1965). The logic of collective action is that in large groups it is rational to expend less effort in the pursuit of things that would benefit the group as a whole. Though the two expressions refer to the same underlying effect, I have adopted "tragedy of the commons" here, both because Hardin's phrase is more evocative and widely known and because in social situations, the combined attention of the group feels more like a natural by-product of social life than a formally created good.

Page 54: **ridiculously easy group-forming** This formulation, by Seb Paquet, a computer scientist at University of Quebec and Montreal, first appeared in 2002 as "Making Group-Forming Ridiculously Easy" (radio.weblogs.com/0110772/2002/10/09.html). The intuition that much of the internet's value comes from its utility as a group-forming tool is often called Reed's law, after David Reed, who described the phenomenon in "That Sneaky Exponential" (www.reed.com/Papers/GFN/reedslaw.html). Reed's law is that "the value of a group-forming network increases exponentially with the number of people in the network," which is to say that value grows even faster for groups than for pairs (as with Metcalfe's law, described above). Paquet amended Reed's law, adding "and in inverse proportion to the effort required to start a group." In other words, the value of a network that allows for group communication will be hindered if it is nevertheless hard to form groups, and helped if it is easy.

## CHAPTER 3: EVERYONE IS A MEDIA OUTLET

Page 58: *Bureaucracy: What Government Agencies Do and Why They Do It*, James Q. Wilson, Basic Books, (1991). Easily the most complete account of the motivations and behaviors of bureaucratic organizations.

Page 60: **mass amateurization** I introduced the formulation "mass amateurization" in an earlier essay, "Weblogs and the Mass Amateurization of Publishing," at shirky.com/writings/weblogs_publishing.html. Charlie Leadbetter, a writer in the United Kingdom, has made similar observations but comes to a different conclusion, calling the effects of peer production a "Pro/Am Revolution," where the work of professionals is increasingly being augmented by that of amateurs in a kind of hybridization. Leadbetter first laid out this argument for Demos, a U.K.-based think tank, in an essay, "The Pro-Am Revolution," at www.demos.co.uk/publications/proameconomy, and later as a downloadable book at www.wethinkthebook.net.

Page 61: **Trent Lott** There is an excellent study of the effect of weblogs on Trent Lott's eventual apology and subsequent resignation at the online journal Gnovis. Called "Parking Lott" (www.gnovisjournal.org/files/Chris-Wright-Parking-Lott.pdf), it documents the absence of the story from the traditional press while it was being widely discussed on weblogs. William O'Keefe's description of the events are at "'Big Media' Meets the 'Bloggers,'" from the Joan Shorenstein Center on the Press, Politics and Public Policy (www.ksg.harvard.edu/presspol/research_publications/case_studies/1731_0.pdf). Ed Sebesta's Anti-Neo-Confederate weblog and list of articles is at newtknight.blogspot.com.

Page 66: **In Praise of Scribes** The printing press, as improved by the invention of movable type, remains the benchmark information revolution. The most complete account of the enormous changes in intellectual, religious, political, and economic life occasioned by increasingly abundant and cheap printed matter is Elizabeth L. Eisenstein's two-volume work *The Printing Press as an Agent of Change*, Cambridge University Press (1979). Eisenstein also has an abridged volume of the same history, *The Printing Revolution in Early Modern Europe*, Cambridge University Press (2005).

Page 75: **Crowdsourcing** Jeff Howe introduced the term "crowdsourcing" in a 2006 article for *Wired* magazine, available at www.wired.com/wired/archive/14.06/crowds.html. Howe is currently at work on a book by the same name and writes a weblog on the subject at crowdsourcing.typepad.com.

## CHAPTER 4: PUBLISH, THEN FILTER

Page 84: **social networking site** After the 2002 success of Friendster, the first widely adopted social networking service, many more were created. Judith Meskill created a list of over three hundred (!) social networking services by 2005, and many more have been created since then. That list, though no longer updated, is at socialsoftware.weblogsinc.com/2005/02/14/home-of-the-social-networking-services-meta-list/.

Two interesting pieces on social networking are: danah boyd's "Identity Production in a Networked Culture: Why Youth Heart MySpace" (transcript of her AAAS talk from 2006 at www.danah.org/papers/AAAS2006.html), describing the forces that led to the success of those services among teens; and an untitled weblog post by Danny O'Brien (www.oblomovka.com/entries/2003/10/13) describing the tensions among public, private, and secret modes of conversation in social media.

Page 94: **Email is such a funny thing** Merlin Mann offered that description of email at "The Strange Allure (and False Hope) of Email Bankruptcy" (www.43folders.com/2007/05/30/email-bankruptcy-2/).

Page 99: **"Conversation is king. Content is just something to talk about."** Cory Doctorow offered that observation in a blog post on BoingBoing.net entitled "Disney Exec: Piracy Is Just a Business Model" (www.boingboing.net/2006/10/10/disney-exec-piracy-i.html).

Page 100: **community of practice** Etienne Wenger first published on this subject in *Communities of Practice: Learning, Meaning and Identity,* Cambridge University Press (1998), and writes more on it (and about social learning generally) at www.ewenger.com.

## CHAPTER 5: PERSONAL MOTIVATION MEETS COLLABORATIVE PRODUCTION

Page 111: **wikis** Wikis are one of the great surprises of the last ten years' worth of work on social tools. While many such tools were simply updates of work done in the 1960s through 1980s, wikis offered a genuinely new pattern of interaction. There are now millions of wikis in operation, both out in public and inside organizations. Ward Cunningham's original wiki is still in operation at c2.com/cgi/wiki. The Wikimedia Foundation, nonprofit parent of Wikipedia, has a number of other wiki-based projects in operation, all listed at wikimedia.org. One of the best descriptions of the history and development of Wikipedia itself is at Marshall Poe's excellent "The Hive," *Atlantic Monthly,* September 2006, and at www.theatlantic.com/doc/200609/wikipedia.

Page 118: **Division of labor is usually associated** If you want to get a sense of the division of labor for Wikipedia, choose any article and look at the top of the page. There, at the edge of the article itself, you will see a set of links. The History link will take you to a page listing all the edits to the article, most recent first. The Talk link will take you to a page where Wikipedians are discussing how the article should be organized and what it should include, including especially conversations about any controversies concerning the content or form of the article. In some cases, the Talk page is longer than the article itself. Both of these exercises are good ways to see the work that goes on behind the scenes.

Page 122: **"Worse Is Better"** The phrase, "The Rise of 'Worse Is Better'" is a section title of Richard P. Gabriel's 1991 essay, "Lisp: Good News, Bad News, How to Win Big" (www.dreamsongs.com/WIB.html). Though most of the essay was addressed to a small community of programmers using the Lisp language, the logic of the "Worse Is Better" argument has spread well beyond that community.

Page 123: **Mermaid Parade graph** This graph is from data I collected from Flickr in June 2005, just after that year's Mermaid Parade. (As an indicator of the astonishing spread of digital photography, at the time of this writing there are nearly thirty thousand photos on Flickr tagged "mermaidparade," a nearly tenfold increase in just two years.) Though I first did the research on Mermaid Parade photos, the subject doesn't matter very much; there is some variation in the steepness of the falloff from the most popular items and the length of the tail of one-off contributors, but the basic power law distribution is stable over most of Flickr (and indeed, over most large social systems.)

Page 124: **power law distribution** A good guide to the ubiquity and interpretive importance of power law distributions in social systems is *Linked: The New Science of Networks*, by Albert-Laszlo Barabasi, Perseus (2002).

Page 126: *The Long Tail: Why the Future of Business Is Selling Less of More*, by Chris Anderson, Hyperion (2006). Anderson, the editor-in-chief of *Wired* magazine, also has a weblog on the subject at thelongtail.com.

Page 129: **fame** I made earlier drafts of these arguments in the essays, "Communities, Audiences, and Scale", www.shirky.com/writings/community_scale.html, and "Why Oprah Won't Talk To You. Ever.", in *Wired* Magazine (August, 2004.)

Page 133: **Yochai Benkler's** *The Wealth of Networks: How Social Production Transforms Markets and Freedom*, Yale University Press (2006) links economics with political and legal theory, sketching out a vision of a world where "commons-based peer production" is allowed to flourish.

Page 136: **Wikipedia deletion and restoration** Martin Wattenberg and Fernanda B. Viégas's work on visualizing the history of Wikipedia edits, "History Flow," is at www.research.ibm.com/visual/projects/history_flow/.

Page 138: **Seigenthaler and essjay controversies** The Wikipedia articles on the controversy surrounding the John Seigenthaler entry (en.wikipedia.org/wiki/John_Seigenthaler_Sr._Wikipedia_biography_controversy) and essjay's faked credentials (en.wikipedia.org/wiki/Essjay_controversy) are surprisingly good, given that one might expect Wikipedians to pull their punches. Nicholas Carr is also worth reading on this subject; Carr, writing at roughtype.com, is the most insightful and incisive of Wikipedia's critics. One of his posts worth reading on the essjay controversy is "Wikipedia's credentialism crisis" (www.roughtype.com/archives/2007/03/wikipedias_cred.php) and

Page 140: **Ise Shrine** Howard Mansfield first noted the linking of the Ise Shrine's method of construction with its failure to win historic designation from UNESCO in *The Same Ax, Twice: Restoration and Renewal in a Throwaway Age*, University Press of New England (2000).

## CHAPTER 6: COLLECTIVE ACTION AND INSTITUTIONAL CHALLENGES

Page 143: **Boston Globe** The Geoghan abuse story was first reported by Matt Carroll, Sacha Pfeiffer, and Michael Rezendes; their stories, as well as other aspects of the abuse scandal, including articles about the Porter case in 1992, have been gathered by the *Boston Globe* in a section called "Spotlight Investigation: Abuse in the Catholic Church" at www.boston.com/globe/spotlight/abuse/.

Page 144: **Voice of the Faithful** James Muller and Charles Kenney lay out the early history of Voice of the Faithful in *Keep the Faith, Change the Church*, Rodale (2004).

Page 150: **Survivors Network of those Abused by Priests** SNAP can be contacted online at www.snapnetwork.org. As of late 2007, they had local chapters in forty-four states, as well as in Canada and Mexico.

Page 157: **end-to-end communication** The idea of end-to-end communication is one of the core design concepts of the internet. The original technical argument is laid out in "End-to-end Arguments in System Design" by Jerome Saltzer, David Reed, and David Clark, available at web.mit.edu/Saltzer/www/publications/endtoend/endtoend.pdf. The argument has been reexpressed in many places; two notable (and notably readable) versions are David Isenberg's "Rise of the Stupid Network" (www.hyperorg.com/misc/stupidnet.html) and "World of Ends: What the Internet Is and How to Stop Mistaking It for Something Else" by Doc Searls and David Weinberger (www.worldofends.com).

Page 157: **the phone company fought bitter legal battles** Telecommunications firms are still fighting to keep control of their customers, a fight that goes back to the landmark Carterfone case. Tom Carter, an entrepreneur, invented a device in the mid-1950s that linked radio and telephone networks. AT&T declared the device illegal; Carter sued and, in 1968, he won. The Federal Communications Commission listed the conditions under which citizens could attach devices to the phone network, a ruling that paved the way for the use of modems and general public access to online services. NPR has a good timeline of the Carterfone decision, and of its continued relevance today, at www.npr.org/templates/story/story.php?storyId=12344564.

## CHAPTER 7: FASTER AND FASTER

Page 161: **Conspiracies are punished separately** Judge Richard Posner. *U.S. v. Wei Min Shi*, 7th Circuit, 02-2241, and at www.projectposner.org/case/2003/317F3d715.

Page 162: **information cascade** "The Dynamics of Informational Cascades: The Monday Demonstrations in Leipzig, East Germany, 1989–91," Susanne Lohmann, *World Politics* 47 (1) October 1994, pp. 42–101.

Page 164: **Flash Mobs** Bill Wasik's description of his launch of flash mobs, and his intended critique of the participants, was published as "My Crowd" in *Harper's* magazine (March 2006) and online at www.harpers.org/archive/2006/03/0080963. Links to photos of several of the Belarusian flash mobs can be found at community.livejournal.com/by_mob/; photos of the *Nasha Niva* protest are at freejul.livejournal.com (both pages are in Belarusian). The events were brought into the English-language blogosphere by Veronica Khokhlova of Global Voices, in "Belarus: Ice Cream–Eating Flash-Mobbers Detained" (www.globalvoicesonline.org/2006/05/15/belarus-ice-cream-eating-flash-mobbers-detained/).

Page 171: *Brave New War: The Next Stage of Terrorism and the End of Globalization*, by John Robb, Wiley (2007). Robb also writes at http://globalguerrillas.typepad.com. Though there is growing agreement among military strategists that social media magnifies the power of "nonstate actors" (all forces on the world stage not tied to countries, including guerrilla and protest movements), there is some disagreement on how to characterize them. For an alternate view to Robb's, read Thomas P. M. Barnett's *The Pentagon's New Map: War and Peace in the Twenty-first Century*, Putnam Adult (2004). Barnett also writes at http://www.thom aspmbarnett.com.

Page 174: *Smart Mobs: The Next Social Revolution*, Howard Rheingold, Tandem (2003).

Page 177: **Kate Hanni** The Coalition for an Airline Passengers' Bill of Rights maintains a weblog at strandedpassengers.blogspot.com, as well as a site with links to background material and the online petition at www.flyersrights.com.

Page 180: **HSBC/Facebook standoff** This was breaking news as the story was going to the printer; the coverage by the U.K. newspaper the *Guardian* was particularly good. The essential details were told in a pair of stories, "Now It's Facebook vs. HSBC," August 25, 2007, at money.guardian.co.uk/creditanddebt/studentfinance/story/0,,2155696,00.html; and "Facebook Campaign Forces HSBC U-turn," August 30, 2007, at money.guardian.co.uk/saving/banks/story/0,,2159132,00.html.

## CHAPTER 8: SOLVING SOCIAL DILEMMAS

Page 190: **Robert Axelrod's** astonishing work, *The Evolution of Cooperation*, Basic Books (1984) is almost single-handedly responsible for turning a simple psychological game into an entire field of research; his book was one of the early examples of real social science being done with computer simulations. Axelrod's follow-up book, *The Complexity of Cooperation: Agent-Based Models of Competition and Collaboration*, Princeton University Press (1997) is an accounting of some of the additional questions of cooperation the original work raised (though it is far more technical than the original work.)

Page 192: *Bowling Alone: The Collapse and Revival of American Community*, Robert D. Putnam, Simon and Schuster (2000).

Page 195: **Meetup** To get a sense of the scope and diversity of the Meetup.com groups, the best place to start is their Browse page at www.meetup.com/browse/.

Page 200: **Club Nexus** "A social network caught in the Web" by Lada A. Adamic, Orkut Buyukkokten, and Eytan Adar, from the online journal First Monday, June 2003, at firstmonday.org/issues/issue8_6/adamic/index.html. Both Adamic and Adar were part of Bernardo A. Huberman's remarkably fecund Information Dynamics Lab at Hewlett Packard (www.hpl.hp.com/research/idl/people/huberman/).

## CHAPTER 9: FITTING OUR TOOLS TO A SMALL WORLD

Page 215: *Small Worlds: The Dynamics of Networks between Order and Complexity*, Duncan Watts, Princeton University Press (1999). *Small Worlds* is Watts's dissertation in book form; he covered the same material with less mathematical density and more real-world examples in *Six Degrees: The Science of a Connected Age*, W.W. Norton and Company (2003).

Page 217: *The Tipping Point: How Little Things Can Make a Big Difference*, Malcolm Gladwell, Little, Brown (2000).

Page 222: **Howard Dean's presidential campaign** The Howard Dean campaign in 2003–2004 was the high-water mark of the use of the internet in national politics, and a number of us were watching it closely (and generally enthusiastically) during that time. Dean's actual performance, once he faced real voters, was so catastrophic that understanding what had happened became a critical task. I wrote two essays on the electoral implosion of the Dean campaign in early 2004: "Is Social Software Bad for the Dean Campaign?" (many.corante.com/archives/2004/01/26/is_social_software_bad_for_the_dean_campaign.php) and "Exiting Deanspace," a reference to a social tool used by the campaign (many.corante.com/archives/2004/02/03/exiting_deanspace.php).

Page 224: **bonding and bridging social capital** Robert D. Putnam followed up his 2000 book *Bowling Alone* with *Better Together: Restoring the American Community*, which he cowrote with Lewis Feldstein and Donald J. Cohen, Simon & Schuster (2003). *Better Together* extends the ideas of bridging and bonding capital in the debate about the decline of social capital in the U.S. context and what to do about it.

Page 224: **social networks and divisions in American class structure** These observations first appeared in danah boyd's essay "Viewing American Class Divisions through Facebook and MySpace" (www.danah.org/papers/essays/ClassDivisions.html). Though boyd is careful to note that she is offering an anecdote of the class divisions manifesting themselves in MySpace and Facebook, rather than a quantitative analysis, her essay has ignited an enormous (and enormously important) discussion of the ways in which our new social tools are bent to the foibles of the users occupying them.

Pages 225–228: **#joiito and #winprog** Internet Relay Chat (IRC) is an unusual social tool in that there is no good general-purpose access to IRC channels through the Web (unlike usenet or mailing lists). IRC requires special software to be downloaded and run on your PC; however, many long-lived IRC groups also maintain webpages. Information about #joiito can be found at joi.ito.com (where else?); information about the #winprog group is at winprog.org.

Page 229: **"The Social Origins of Good Ideas,"** Ronald S.Burt, *American Journal of Sociology* (2004) and at web.mit.edu/sorensen/www/SOGI.pdf.

## CHAPTER 10: FAILURE FOR FREE

Page 233: **Failure for Free** This argument first appeared in *Harvard Business Review* (February 2007) under the title "In Defense of 'Ready. Fire. Aim.'"

Page 240: **Open source software** There is an enormous amount of material trying to explain open source software, most of it mediocre. The most rigorous overview on the topic is Steven Weber's *The Success of open source*, Harvard University Press (2004), which provides a detailed description of the development of Linux, as well as an excellent theoretical analysis of what makes open source projects work.

Page 242: **"The Cathedral and the Bazaar"** As noted for chapter 1, Eric Raymond's seminal 1998 essay on open source software, "The Cathedral and the Bazaar," is at catb.org/~esr/writings/cathedral-bazaar/. Raymond's writings on software and other topics is at www.catb.org/~esr/writings/.

Page 244: **Sourceforge** Sourceforge, at sourceforge.net, is the largest repository of open source projects; the list of projects sorted by "activity" (a composite metric of various different gauges of programmer and user engagement) is at sourceforge. net/top/mostactive.php.

Page 247: *Wikinomics: How Mass Collaboration Changes Everything*, Don Tapscott and Anthony D. Williams, Portfolio (2006).

Page 250: **Nick McGrath** McGrath's comments can be found in Robert Jaques's 2005 VNUnet article, "Linux security is a 'myth,' claims Microsoft," at www.vnunet. com/vnunet/news/2126615/linux-security-myth-claims-microsoft.

Page 253: **Groklaw** Groklaw's mission statement is at www.groklaw.net/staticpages/ index.php?page=20040923045054130; in it, she notes that "we are applying open-source principles to research to the extent that they apply. Our community includes those with a technical background and others with legal and paralegal training, as well as journalists, educators, and many end users who care enough about their operating system of choice to work to defend it." SCO, the company that has so far unsuccessfully tried to sue IBM, became so frustrated with the work Groklaw was doing that they accused Pamela Jones, the founder, of being funded by IBM. Jones categorically denied this charge, and called out the idea of nonfinancial motivation in her response to SCO: "Groklaw is a labor of love. SCO seems to find it hard to believe that I would do this as a volunteer. But I do. They don't understand wanting to pool knowledge period, being a bit old-fashioned in their thinking." ("Letter to the Editor: No IBM-Groklaw Connection," news.zdnet.com/2100-9595_22-5170485.html).

Page 254: **SARS** Dr. Yang Huanming's lament about the obstacles to their work can be found in translation at the YaleGlobal journal, in "Chinese Scientists Say SARS Efforts Stymied by Organizational Obstacles" (yaleglobal.yale.edu/display. article?id=1745). Martin Enserink has a broader review of the Chinese performance in "SARS in China: China's Missed Chance" *Science* 301 (5631), July 18, 2003, and at www.sciencemag.org/cgi/content/full/301/5631/294).

## CHAPTER 11: PROMISE, TOOL, BARGAIN

Page 267: *The Wisdom of Crowds: Why the Many Are Smarter Than the Few and How Collective Wisdom Shapes Business, Economies, Societies and Nations*, James Surowiecki (Doubleday, 2004)

Page 276: **equality matching** The idea of equality matching (and the other listed forms of social participation) come from Alan Page Fiske's *Structures of Social Life: The Four Elementary Forms of Human Relations: Communal Sharing, Authority Ranking, Equality Matching, Market Pricing*, Free Press (1991). Fiske also provides a brief account of these ideas in "Human Sociality" at (www.sscnet.ucla.edu/anthro/fac-ulty/fiske/relmodov.htm).

Page 281: **"Sluggy Freelance"** Jessica Hammer was a graduate student in 2002 at the Interactive Telecommunications Program, NYU, when she did this research.

Page 281: **Usenet** Usenet was one of three great global experiments in social tools prior to the invention of the Web. (The other two were e-mail discussion lists and online communities such as the WELL and ECHO.) At the height of its popularity, in 1994, usenet was at the core of most users' experience of the internet. (Though still in operation, its subsequent decline was a result both of a shift to the Web and because it had no built-in defenses against the tragedy of the commons that is spam.) Usenet is organized into "newsgroups" (in quotes because most are not devoted to anything that could be called news), loosely categorized by topic (comp. lang.perl, for example, is about the Perl computer language.). The easiest way to get to these newsgroups is through groups.google.com, which provides a Web-based interface to the groups.

Page 282: **civic bicycle programs** Interestingly, many accounts of the failure of the original White Bicycle program include an unsubstantiated accusation that the bicycles were confiscated or thrown in the canals by the police. These stories create the sense that uncontrolled bike-sharing would have succeeded but for this intervention by the authorities; such stories, however, are hard to make sense of in light of the collapse of uncontrolled programs in subsequent eras. You can get some sense of the universality of the problem of theft by looking at antitheft instructions at contemporary community bike Web sites like ibike (www.ibike. org/en couragement/freebike-issues.htm#TRACKING).

Pages 287–288: **sending nuts and flowers** The campaign to save the TV show *Jericho*, including sending CBS nuts, was coordinated at jericholives.com. The protest was an echo of General Anthony McAuliffe's one-word reply—"Nuts!"—to a German request that U.S. forces surrender during the Battle of the Bulge in World War II. (Amusingly, NutsOnline also hosted a page on the campaign, at www.nutsonline. com/jericho.) The antiwar flower protest in Michigan was a way of doing "something positive to deliver our message," as one protester put it ("Flowers Used to Protest War," www.statenews.com/index.php/article/2006/04/flowers_used_to_ protest). Similarly, the flowers sent to the U.S. State Department were often referred to as Ghandigiri, which is to say "in the spirit of Mahatma Gandhi" ("Say It with Flowers: Gandhigiri for US Green Cards," in.news.yahoo.com/070710/48/ 6hwnn.html). In all these cases, the delivery of actual objects did triple duty: the physical delivery increased attention, the nature of the object underlined the message (opposition with the nuts, nonviolence with the flowers), and the cost of sending the object communicated real commitment on the part of the sender.

Page 290: **Digg Revolt** Kevin Rose made his remarks on the official Digg weblog in "Digg This: 09-f9-11-02-9d-74-e3-5b-d8-41-56-c5-63-56-88-c0," at blog.digg. com/?p=74.

## EPILOGUE

Page 304: **lump of labor fallacy** The lump of labor fallacy is well described (and eviscerated) by Paul Krugman in "The Accidental Theorist" (web.mit.edu/krugman/ www/hotdog.html).

Page 309: *Sicko* **audience** Josh Tyler wrote about this in "*Sicko* Spurs Audiences into Action" (www.cinemablend.com/new/Sicko-Spurs-Audiences-Into-Action-5639. html).

# INDEX

Abd El Fattah, Alaa, 184–186, 268
Adamic, Lada, 200
airline passengers, 175–179
Akinola, Archbishop Peter, 155
America Online, 134–135
American Airlines, 177–178
Anderson, Chris, 126
Anderson, Philip, 28
AT&T, 194, 195, 256–258
audience, the former, 7
audiences, 10, 84–90
average (mean), 126, 127, 213
Axelrod, Robert, 190–191

bargains, role in groups, 260, 261,
    270–277
Barlow, John Perry, 194
Belarus, 23, 166–171
bell curve distributions, 126–127
Benkler, Yochai, 133
Berners-Lee, Sir Tim, 158
Birthday Paradox, 25–26, 27, 41, 215, 267
blitzkrieg, 172–173
Blogger (product), 182
bloggers. *See* weblogs
Boing Boing, 72, 88
Bomis company, 109–110
bonding capital, 222–224, 227
*Boston Globe*, 143, 144, 147, 148, 149–
    150, 153, 159

bowling, 191–192
boyd, danah, 224–225
Bradner, Scott, 77
bridging capital, 222, 225–228, 229,
    230–231
broadcast media, 86–88, 89, 90–94,
    95, 98–99, 106–107
Brooks, Fred, 29
Brown, John Seely, 100
*Buffy the Vampire Slayer* (TV show),
    269–270, 281, 284
bulletin boards, online, 4–5, 12, 13,
    203–206, 269–270, 281, 284. *See
    also* webpages
Burroughs, William S., 99
Burt, Ronald, 229–230

camera phones, 74, 169
Carvin, Andy, 167
Catholic Church, 19, 23, 43, 67, 143–
    160, 207, 262
cell phones. *See* mobile phones
Chandler, Alfred, 40
channels, IRC, 225–226, 227, 228
Chirapongse, Alisara, 37, 89
Chui, Howard, 228–229
Clohessy, David, 150–151
Coase, Ronald, 30, 42–43, 44, 45, 47,
    48, 130, 207
collaborative production, 50–51, 109, 143

collective action
Catholic priest scandal as example,
143–160
*vs.* collaborative production, 51, 143
flash mobs as, 164–171
*vs.* individual action, 161–164
overview, 47–54
rapid and simple group formation,
151–153
removing obstacles, 153–156, 159
sharing information, 148–151
Tragedy of the Commons example,
51–53, 135, 137, 190, 275
as type of group action, 51–53
communications media, 86–88, 89, 95,
98–99, 106–107
communications tools. *See also* social
tools
*vs.* broadcast tools, 99
many-to-many, 87, 107, 156, 157–158
modern, 20–21, 23, 48, 77, 87
negative effects, 210
now *vs.* then, 77, 157–158
one-to-many, 87, 158
one-to-one, 86, 87, 158
as substitute for travel, 194–195
technological *vs.* social change, 105–
106
communities, 85, 89, 100–102, 103,
228, 277–279
consumers, use of social tools, 78–79,
179–182
conversation, 50, 86, 95, 99
Cool, Jeannie, 227
cooperation, as type of group action,
49–51
corporations. *See* Microsoft; organizations
cost factor, 18, 31. *See also* transaction
costs
Craigslist, 56
Crowley, Dennis, 218
Cunningham, Ward, 111–112, 117
cyberspace, 194, 195–196, 198

Dasburg, John, 176–177
Dean, Howard, 165, 222–223, 224,
288

del.icio.us service, 263
Digg, 3, 290–291
discussion groups, 10, 256–258, 269–
270, 281, 284. *See also* mailing
lists; weblogs
distributions, power law *vs.* bell curve,
126–127. *See also* power law distributions
Doctorow, Cory, 99
Dodgeball, 218–220, 228
Duguid, Paul, 100
Dyson, Esther, 96

e-mail. *See also* mailing lists
as asynchronous, 157
flash mobs and, 165
as form of publishing, 76
as many-to-many communication
pattern, 87, 107, 156, 157–158
scale issue, 94–95
sharing news stories, 148–151
as social tool, 11–12, 20, 76, 156–157,
160, 270, 292
stolen Sidekick and, 1, 2, 3, 17, 20
as tool for lobbying Congress, 287
transaction costs, 287
eBay, 56, 283–284
Edyvean, Bishop Walter J., 145, 155
Egyptian activists, 184–186, 268
80/20 rule, 125–126, 250, 252, 279
elections, and blogging, 210
Encarta, 254, 289
Enron, 76
Episcopalian Church, 154–155
Estrada, Joseph, 174, 175

Facebook, 84, 180–181, 219, 225
Fake, Caterina, 264–265
Falun Gong, 174, 175
fame, 91–92, 93, 94, 95, 96
FAQs, 269
filtering, 96–98, 249
Fiske, Alan Page, 276
fitness landscape, 247–248, 254
flash mobs, 164–171, 272–273
Flickr
acquired by Yahoo, 289–290
Belarus flash mob pictures, 167

Black and White Maniacs, 275–276, 281

high dynamic range (HDR) photos, 99–100

how it works, 32–33

participation imbalance, 123, 124, 125, 127

power law distribution and, 124–125, 127, 237

promise concept, 264–265

role in breaking news stories, 34–37, 38, 39, 48, 66

role in Coney Island Mermaid Parade, 31–34, 38, 169

as sharing platform, 49, 51, 101, 102–103

significance, 38–39, 46

user-generated content and, 82, 83

Flyers Rights group, 263, 268. *See also* airline passengers

former audience, 7, 10

Free Software Foundation (FSF), 240–241, 242, 243

freedom, 305–306

friend-of-a-friend (FOAF) networking, 219–220

Friends of O'Reilly (FOO) Camp, 229

Gabriel, Richard, 122

Genbank, 255

Genome Sciences Centre (GSC), 255

Geoghan, Father John, 143–144, 147, 149–150

Gibson, William, 165, 194

Gillmor, Dan, 7, 72

Gladwell, Malcolm, 217

GNU Public License (GPL), 238, 241, 243, 273–274

Goldcorp mining, 247–249

good ideas, 229–232

Google, 49, 182

Groklaw, 252–253

group action

collapse of institutional barriers to, 21–24, 31

collective action as, 51–53

cooperation as, 49–51

coordinating, 31, 45–47

sharing as, 49, 148–151, 275, 276–277

stolen Sidekick and, 7

groups

advantage of "ridiculously easy group-forming," 54

complexity, 14–17, 25–29, 278

ease of formation, 18–19

large *vs.* small, 267–268

latent, 38, 102, 169, 206, 210–211

net value argument, 303–305

paradox, 263

political value, 305–318

role of bargain, 260, 261, 270–277

role of promise, 260, 261–265

role of social tool, 260, 261, 265–270

Gutenberg, Johannes, 67, 153

Guttman, Evan, 1–14, 17, 20, 48, 63, 275

Hackman, Richard, 31

Hammer, Jessica, 281

Hanni, Kate, 177–179, 181, 263

Hardin, Garrett, 51, 53

Heiferman, Scott, 193–194, 196–197, 199

heralding, 227

Hewlett-Packard, 100, 243

hierarchical organizations, 30, 39–46, 153, 154

high dynamic range (HDR) photos, 99–100

Holmes, Oliver Wendell, 274, 275

homeostasis, 283–285

homophily, 213, 224

Honecker, Erich, 162

Howard Forums, 228–229, 284

Howe, Jeff, 75

HSBC bank, 180–181

Huanming, Yang, 255

Huberman, Bernardo A., 329

IBM, 136, 243, 252–253, 279

Indian Ocean tsunami, 36, 38, 48, 66

information cascades, 162–164

instant messaging, 20, 50, 87, 183, 226
Instapundit.com, 62–63, 93
institutional dilemma, 19–21, 31, 39,
    41–42, 46, 60–61, 120–121
institutions. See organizations
IRC (internet relay chat), 225–228
Ise Shrine, Japan, 140, 141
iStockphoto, 75
Ito, Joi, 225–228

Jaeggli, Erika, 203–204, 205
Jardin, Xeni, 72
Jetblue, 178
jibot, 226–227
Jones, Pamela, 253
journalistic privilege. See also news
    business
Joy, Bill, 254

Keen, Andrew, 209
Kinsley, Michael, 286
Krucoff, Andy, 219–220

latent groups, 38, 102, 169, 206, 210–
    211
Law, Cardinal Bernard F., 144, 145, 146,
    147, 150
Leipzig, East Germany, 161–164
Lindbergh, Charles, 93
Linux, 238–243, 244, 250–252, 253–
    254, 262, 266, 273, 281. See also
    open source software
literacy, 67, 78, 79, 304
LiveJournal
    Belarus flash mob, 166–167, 170
    I Love My Boyfriend group, 285
    Meetup group, 197, 199, 235
    promise concept, 264
    user-generated content and, 81–82,
        84, 85
    as weblog tool, 89–90
Lohmann, Susanne, 162
London Transport bombings, 34–35, 38,
    45, 116–117
Los Angeles Times, 137, 283, 285–286
Lott, Trent, 61–66
Lukashenko, Alexander, 166–171
Luther, Martin, 67, 154, 156

Mackinnon, Rebecca, 72
Madrid train bombing, 174
Mahmoud, Abdel Monem, 185–186
mailing lists, 10, 101–102, 124, 127, 237,
    285. See also discussion groups; e-
    mail
Mann, Merlin, 94–95
Manutius, Aldus, 318–320
many-to-many communications tools,
    87, 107, 156, 157–158
mass amateurization, 60–66, 98, 112,
    122–130, 209
McCallum, David, 41–42, 46
McGrath, Nick, 250–251, 253
mean (average), 126, 127, 213
media industry. See also news business
    broadcast media vs. communications
        media, 86–88, 89, 95, 98–99
    mass amateurization and, 60–61,
        64–66, 209
    revolutionary changes, 107–108
Meetup
    convening power, 198
    Dean campaign and, 288–289
    as example of Small World network,
        218, 280
    failure and, 233–237
    fitness landscape and, 248
    how groups form, 233–237
    launching, 195–197
    most active groups, 197–200
    social capital and, 224
    Stay at Home Moms (SAHM), 200–
        202, 233, 234, 235, 280
Mermaid Parade, Coney Island, 31–34,
    38, 169
Meyer, Chris, 195
Microsoft, 19, 238, 239, 243, 250, 251–
    252, 254, 289
Miller, Judith, 70, 73
Misilim, Marion, 285
mobile phones
    as digital cameras, 74, 169
    Dodgeball service, 218–220
    Howard Forums and, 228–229, 284
    as revolutionary change, 106–107, 160
    shift away from advance planning,
        174–175

as social tools, 301
stolen Sidekick story, 1–14
Twitter service, 183–187
Moore, Michael, 309
movable type, 66–69, 79, 106, 319
MoveOn.org, 286, 287, 288
Muller, James, 144
Murad, Abdel Fatah, 185
MySpace
    California school boycott, 221–222
    as example of Small World network,
        218, 219, 221–222, 279–280
    vs. Facebook, 225
    participation imbalance, 127
    promise concept, 264
    significance, 11–12
    stolen Sidekick and, 3, 4, 6, 9
    user-generated content and, 82, 84

Nash equilibrium, 190
net value argument, 303–304
New York City Police Department
        (NYPD), 5–6, 8, 9, 10–11, 13
news business. See also media industry
    internet and, 56–57, 59–60
    Lott/Thurmond example, 61–66
    as profession, 57–60, 64
    sharing news stories, 148–151
    trustworthiness issue, 65–66
    USA Today threat, 55–56
Northwest Airlines, 175–177
Nupedia, 109–111, 112, 113, 121, 139
NYPD. See New York City Police
        Department

O'Keefe, William, 62
Omidyat, Pierre, 283
one-to-many communications tools, 87,
        158
one-to-one communications tools, 86,
        87, 158
open source software, 18, 240, 242,
        243–247, 249, 251, 252, 253, 254,
        255, 258
O'Reilly, Tim, 17, 229
organizations. See also institutional di-
        lemma
    Catholic Church, 153–156

collapse of barriers to group action,
        21–24
diagramming hierarchies, 39–42
hierarchies and transaction costs,
        42–46
railroads, 40–42
running, 29–31
vs. self-assembly, 47–54
symmetrical participation, 107–108
transaction costs in, 29–30, 248–252

Paquet, Seb, 54
Pareto, Vilfredo, 126
Partido Popular, 174, 175
Perl programming language, 256–258
phones. See mobile phones; telephones
photographers, 31–39, 74–75, 99–100,
        169. See also Flickr
photos, high dynamic range (HDR),
        99–100
political expression
    Belarus protest, 166–171
    Dean presidential campaign, 165
    flash mobs and, 164–171
    in Leipzig, 161–164
    in Philippines, 174, 175
    protest against Putin, 165–166
Porter, Rev. James R., 146–147, 148, 149
Posner, Richard, 161
power law distributions, 123–130, 217,
        237, 248–252
printing press, 69, 73, 77, 106, 153,
        303, 304, 319. See also movable
        type
Prisoners' Dilemma, 189–190, 191
Pro-Ana movement, 204–208, 210,
        224, 262
professionals
    in advertising agencies, 209
    journalists as, 70–74
    medieval scribes as, 66–69
    photographers as, 74–75
    publishing and, 77–80
    scarcity concept, 57, 58–59, 64, 67,
        73, 74, 76–77, 79, 83, 98
    self-definition, 57–60
    in stolen Sidekick story, 17
promise, role in groups, 260, 261–265

publishing
    *vs.* filtering, 96–98, 249
    globally accessible, 171
    now *vs.* then, 77–80
    stolen Sidekick and, 9
Putin, Vladimir, 165–166
Putnam, Robert, 192, 193

railroad management, 40–42
Rainert, Alex, 218
Raymond, Eric, 18, 242
Reynolds, Glenn, 62–63, 93
Rheingold, Howard, 174
Ripley, Andy, 181
Robb, John, 171
Robinson, Gene, 154
Rose, Kevin, 290, 291–292
Ruiz, Victor, 226–227

Sanger, Larry, 109, 110, 111, 112, 113,
    115–116, 277
Santa Cruz Organization (SCO), 252–
    253
SARS (Sudden Acute Respiratory
    Syndrome), 254–255
scarcity, impact of social tools on con-
    cept, 57, 58–59, 64, 67, 73, 74, 76–
    77, 79, 83, 98
Schachter, Joshua, 263
scribes, medieval, 66–69, 79, 304
Sebesta, Ed, 63
self-help movement, 207–208
self-publishing, 77–80
shadow of the future, 191, 192–193
sharing, as type of group action, 49,
    148–151, 275, 276–277
Shinawara, Thaksin, 37
*Sicko* (movie), 309–310
Sidekick phone story, 1–14
Siegenthaler, John, Sr., 138
Site Specific, 256
Sites, Kevin, 72
six degrees of separation, 214, 218
Sluggy Freelance, 281
Small World networks
    as amplifiers, 221
    characteristics, 215–217
    examples, 218–220

    as filters, 221
    power law distribution and, 217
    promise concept, 264
    scale issues, 215, 220–225, 279
    small group advantage, 267
    social capital in, 222, 224
Smith, Adam, 79
social capital
    bonding *vs.* bridging, 222, 224
    declining, 193–194
    defined, 192, 222
    Meetup example, 195–202
    reinvigorating creation, 193–195
    relationship to good ideas, 230–232
    in Small World networks, 222, 224
social loss, 209–211
social networks, 212–214. *See also*
    Facebook; LiveJournal; MySpace;
    Small World networks; Xanga
social participation modes, 276–277
social tools. *See also* Flickr; Meetup; so-
    cial networks; Twitter; weblogs
    advantage of "ridiculously easy
        group-forming," 54
    audience for user-generated content,
        84–90
    becoming ubiquitous, 105–106, 320,
        321
    communities of practice and, 100–
        102, 103
    controlling, 300–302
    corresponding social patterns, 48–
        49, 54
    e-mail as, 11–12, 20, 76, 156–157,
        160, 270, 292
    flash mobs, 164–171
    goodness of fit, 265–266
    mobile phones as, 301
    net value argument, 303–304
    participation imbalance, 124, 127,
        128–130
    photo-sharing websites, 31–39, 74–
        75, 99–100, 169
    political value, 305–318
    as response to institutional dilemma,
        20–21
    as revolutionary, 107–108, 160
    role in groups, 260, 261, 265–270